Vein Pattern Recognition

A Privacy-Enhancing Biometric

Vein Pattern Recognition

A Privacy-Enhancing Biometric

Chuck Wilson

CRC Press is an imprint of the
Taylor & Francis Group, an **informa** business

CRC Press
Taylor & Francis Group
6000 Broken Sound Parkway NW, Suite 300
Boca Raton, FL 33487-2742

First issued in paperback 2017

ISBN 13: 978-1-138-11531-6 (pbk)
ISBN 13: 978-1-4398-2137-4 (hbk)

Library of Congress Cataloging-in-Publication Data

Wilson, Chuck.
Vein pattern recognition : a privacy-enhancing biometric / Chuck Wilson.
p. cm.
"A CRC title."
Includes bibliographical references and index.
ISBN 978-1-4398-2137-4 (hardcover : alk. paper)
1. Biometric identification. 2. Blood-vessels. I. Title.

TK7882.B56W55 2010
658.4'73--dc22 2009047362

Visit the Taylor & Francis Web site at
http://www.taylorandfrancis.com

and the CRC Press Web site at
http://www.crcpress.com

DEDICATION

This book is dedicated to Masakazu Hamada and Katsuhiro Kou, whose leadership and integrity have been an inspiration.

CONTENTS

FOREWORD

Although they have become a favorite prop, used in science fiction movies and spy novels, the technology and systems behind biometric authentication are not well understood by most people. We interact with biometric-enabled systems every day, sometimes without even realizing it. Because our lives revolve around information, biometric identification and authentication are among the most important technologies of the 21st century, yet most business users do not truly understand biometrics. This book explains the use of vein pattern recognition and other biometric methods, how they work, and how they are used to protect sensitive information and access to critical facilities in the real world.

As we move forward and face the challenges of the 21st century, security of personal information will continue to be foremost in our thoughts. Our lives are already heavily dependent on secure information flow since nearly everything we touch has some connection to information processing. Bank accounts, transportation systems, mobile phones, and computers are all connected to networks, allowing them to serve our needs. The first question that any of these systems asks us is: "Who are you?" The ability to determine our true identity is critical to ensure the protection of both our personal information and the networks that underpin the digital fabric of society. Without personal identification numbers (PINs), passwords, tokens, and now biometrics, such systems would not be able to reliably know who we are and act to protect our interests. Answering the fundamental "Who are you?" question has implications on us whether we understand them or not. Networks often have subtle vulnerabilities that could allow malicious parties to acquire our most precious possession—our identity—and use it for their own purposes. Although biometric identification is not the only way to protect our identity in the digital world, it represents a huge step forward, if implemented correctly.

Biometric systems protect our digital persona, and they can enhance our privacy. As more of our social network becomes digitized and accessible online, we will face new challenges associated with the protection of our privacy. Even with strong identification and authentication methods, the privacy of our transactions must be protected so that we can choose to disclose our activities at an appropriate level, depending on the context. Biometric identification methods, combined with other traditional means such as smart cards, can provide a mechanism that is empowering to the individual, while making it convenient for us to do the important things without unwarranted disclosures.

The U.S. Federal Bureau of Investigation's Automated Fingerprint Identification System is a key tool used by law enforcement to investigate crimes. Fingerprints have been an important forensic tool for identifying people since the early 20th century. But as a means of identifying people to computers, they often fall short of the full requirement. Their shortcomings are not obvious at first glance. However, we often forget that we leave behind subtle traces of who we are, hundreds of times every day each time we touch something. Nobody would willingly leave a copy of his or her house key lying around or place a copy of a government-issued identification card on each object he or she touches. Whether we realize it or not, that is exactly what we do when we rely on fingerprints as a sole means of identity verification. However, recent advances in technology make it possible for us to use our physical body as a token for such interactions without leaving traces behind in the physical world.

Some biometric scanners have become almost as inexpensive as a computer mouse, opening new possibilities for our identity to be assumed by fraudsters intent on impersonating us. However, recent advances in vein recognition systems mean that we will soon have a new way to access the digital world as well as the physical one. Since our veins are invisible to the naked eye and embedded within our bodies, near-infrared finger vein scanning technology presents an opportunity to use the unique tokens that we each carry around with us to gain access to desired systems. This presents much less possibility of compromise in leaving latent traces everywhere, as we do with fingerprints. This is highly significant since it means that we can retain the convenience of using our fingers and hands as keys to information systems, without leaving inappropriate copies behind. There are other advantages and many promising applications of this technology outlined in this book.

Vein Pattern Recognition: A Privacy-Enhancing Biometric provides a clear road map for the past, present, and future of biometrics as they are commonly used in systems we interact with every day. As we move toward the future, they provide a way to increase privacy, convenience, and security all at the same time. These seemingly contrasting objectives are the subject of widespread debate and discussion in government and in the press. However, people involved in the discussion may not fully understand these technologies in any great detail. Even fewer among us can predict where they are going.

Chuck Wilson is one of the few exceptions to this rule. His book provides a practical guide to biometrics in a clear and easy to understand form, even for nontechnicians. This thoroughly researched book covers the main technologies in use today and explains the principles of operation and appropriate uses of each. For the technically oriented reader, it provides significant details for each system. However, it is not simply a technical reference, but also a guide for the practitioner or even the average person who is interested in understanding the field. It explains the reality of the science behind the science fiction that we have all seen and strips away the mythology surrounding these important tools.

This book provides an objective comparison of the different biometric methods in common use today including fingerprint; eye, face, and voice recognition; and even dynamic signature verification. These automated tools each have their place and their limitations. This book serves as a great guide to compare them and to select the right combination of technologies for specific applications. There is a wealth of knowledge about how biometrics has been used in systems we interact with every day, ranging from personal computer access to automated teller machines (ATMs) and physical access systems. These cases are complemented by examples that describe the processes underlying the technology. Heightened security must coexist with the preservation of consumer privacy if our society is to move forward.

Understanding the history and motivation behind these technologies provides insight into their appropriate uses, in the present and in the future. It is also important to understand the underlying processes that are made more effective through the incorporation of biometric identification. Since this book does not presume that you are an expert on such technologies, it provides a solid starting point for anyone interested in pursuing them more deeply.

Vein pattern recognition technologies will become more widespread in use and acceptance by the general public over the next several years and the nature of information systems will become more ubiquitous. The many innovations by Hitachi, Fujitsu, Bionics, Identica, Sony, and others will allow us to live in a world in which information technology will simplify our lives without sacrificing convenience or privacy. These systems are already commonplace in leading societies in Asia because the public perceives the need for security and risk management as fundamental. Just like the amazing technologies such as the now near-ubiquitous mobile phone, which quickly gained public acceptance and user demand, leading institutions will learn and adapt vein pattern recognition and other advanced biometrics to global lifestyles. As these systems emerge, those individuals and institutions who know the difference between science and science fiction in the application of biometrics to daily life will benefit personally and possess an advantage over those who do not.

A fundamental debate resolved by this book is whether these technologies are practical, usable, and helpful in today's globally connected digital society. There is no doubt that they are ready and will begin to impact us in positive ways sooner than most people believe. When that happens, we will be one step closer to the dream of ubiquitous information systems that will serve and empower us all.

As famed science fiction author William Gibson once said, "The future is already here—it's just not evenly distributed." Read this book and get a glimpse of how bright the future can be if we are all empowered rather than encumbered by technology.

Kevin R. Walsh
Senior Vice President R&D
Oracle Corporation Asia Pacific

PREFACE

In this newly complicated world of terrorism, identity theft, and rampant consumer fraud, biometrics has been heralded as a key technology for identity management, and hence security. Never before has identity management been so important. Governments and enterprises of all sizes have become much more vigilant regarding security. There is always a need to reexamine and potentially improve security, and biometrics is attracting growing interest as fraud increases and the conventional authentication methods—PINs, passwords, and identity cards—prove inadequate to counter the growing threats. Biometric tools have become prominent differentiators for multiple applications in a variety of markets. The use of biometrics offers no panacea to completely remedy society's threats, and it provides no guarantee against terrorist activities. However, biometric technologies remain a critically important component of the total solution.

The biometric authentication market has emerged and is expanding at an increasing rate. It has sustained a compound annual growth rate (CAGR) of approximately 30% for the past 3 years, and it is expected to sustain a strong growth rate for the next decade.

Biometric systems are proliferating. The diversity of the various modalities and the many false claims of their promoters and detractors alike have somewhat clouded the market with at best some misinformation and at worst a public concern that this new technology is somehow menacing and will restrict freedoms. Unfortunately, many of the key benefits of biometrics have become obfuscated due to unfortunate sensationalism and myths that have surrounded biometric solutions.

Biometric technologies vary in capability, performance, and reliability. The success of a given biometric modality depends not only on the effectiveness of the technology and its implementation, but also on the total security solution for which any biometric system comprises only a part.

The next several years will be exciting for the biometric market. We can expect increased user acceptance and demand as biometrics continue to become more user friendly and more reliable. Improved technology and biometric need are converging. There should be significant growth in each of the various biometric modalities, as well as in multimodal biometrics. We can expect to see more standards surface, enabling greater interoperability. Moreover, with the greater demand on biometrics in everyday life, governments are expected to enact statutes that help administer biometric solutions while maintaining privacy and legal support. Indeed, it has been the use of biometric solutions by government agencies and by mainstream industries such as banking and health care that has increased public awareness and acceptance of the technology.

As more biometric systems are adapted to customer-operating environments, as prices for biometric solutions continue to fall, and as their performance improves, the business case for biometric solutions will gain ascendancy. Our use and dependency on biometrics will continue to expand. Against an estimated $2.8 billion market in 2007, technology consultancy New York City–based International Biometric Group (IBG) has projected that the 2010 biometric market will exceed $5.7 billion. Another market researcher, Acuity Market Intelligence, based in Louisville, Colorado, released its estimates in 2009 and forecasts biometric revenues to exceed $11 billion by 2017. Acuity reported that the growing reliance on digital transactions and the inevitability of available broadband access will require a level of authentication available only through the use of biometrics. Moreover, today's primary biometrics user base—national defense, law enforcement, intelligence, and border control—may soon be eclipsed by the preponderance of commercial applications.

Vein pattern recognition (VPR) technology is emerging and rapidly gaining market share. Also known as vascular pattern recognition, the technology has already achieved a high-profile acceptance at major financial institutions, especially in Japan where most banks employ this type of biometric identifier. Schools, apartment buildings, and public access facilities such as airports and seaports are using vein pattern recognition technology for physical access control. Enterprises of all sizes are using this same technology for time and attendance and logical access control applications. Membership and accountability applications are entering a phase of hypergrowth.

What are the growth drivers for vein pattern recognition? To some extent, it is globalization that has engendered heightened interest in biometrics in general and in vein pattern recognition specifically. The tragic events of 9/11 transformed our mindset about terrorism for the entire world. Globalization requires trusted transactions among remote trading partners and dependable information exchange in a world fraught with high risk amid emerging opportunities. Accompanying the global concerns are specific regional requirements. It is interesting that the current five key developers of vein pattern recognition technologies—Bionics, Fujitsu, Hitachi, Sony, and Techsphere—are Asian companies.

There are a variety of reasons why vein pattern technology is capturing market share so rapidly, but we could probably distill the rationale for its success to three key developments. First, vein pattern technology is more accurate than many other biometric modalities. Second, it offers much greater resistance to spoofing. This is especially timely given the fascination of the press to several highly visible circumventions of some fingerprint technologies. Finally, and perhaps the most important reason is its people focus. Vein pattern technology is easy to use, has an exceptionally high usability rate, is hygienic, is easily adaptable, and is privacy enhancing. This combination of benefits is what has propelled vein pattern technology to enormous growth. Clearly, vein pattern technology is here to stay, and it is gaining momentum. Moreover, it is being marketed aggressively due to high user acceptance and the backing of major corporations on a global scale. I wrote this book to share some thoughts about this most intriguing and fastest-growing biometric modality.

ACKNOWLEDGMENTS

There is no claim of self-sufficiency in writing this book. A number of talented individuals helped me to research topics, develop materials, proofread, edit, and illustrate. My deepest appreciation to the following individuals:

Jerry Byrnes
Jac Cornett
Francine Foster
Christina Franco
Miguel Gonzalez
Steve Hollingsworth
Lew Iadarola
Michiaki Kawagishi
Carola Kessler
Susan Larsen
John Liddell
Masahiro Mimura, Ph.D.
Shimon K. Modi, Ph.D.
Idan Shoham
Roque Solis
Samir Tamer
Terry Wheeler

And special thanks to Mitsutoshi Himaga, Ph.D., whose technical guidance was indispensible to the writing of this book, and to Greg Freed for his firm and consistent editing of this book.

ABOUT THE AUTHOR

Chuck Wilson has worked in the information technology (IT) industry for more than 30 years. He worked in the card processing industry for more than 20 years, and has been researching and writing about smart cards and biometrics throughout this decade. Wilson spent 12 years with Electronic Data Systems (EDS) managing payment services and electronic benefits transfer (EBT) businesses. Wilson was Senior Vice President of CardSystems Solutions Inc., in Addison, Texas, where he led the development of emerging payment products. He was Senior Director at Hitachi America where he headed up the Hitachi Security Solutions business in North America, focusing on biometrics and smart card solutions. Today, Wilson manages and directs the Identity Verification business practice for ii2P, based in Southlake, Texas.

In June 2001, Wilson's first book, *Get Smart* (Mullaney), was published regarding the emergence of smart cards in the United States and their pivotal roles in electronic commerce.

Wilson has met with an extensive variety of biometric manufacturers, service providers, and system integrators in North America and in Asia, and has developed a firsthand understanding of biometric systems. He has been involved in a number of vein pattern biometric initiatives throughout North America. As a member of the Institute of Electrical and Electronics Engineers (IEEE) Certified Biometrics Professional™ Examination Committee, Wilson contributed to the development of IEEE's new professional certification program for the biometrics industry, creating the program's first standardized certification examination.

Wilson received his BA and MA degrees from Ohio State University, and his MBA from Memphis State University, Tennessee. He is a recognized speaker at security and e-commerce seminars and trade shows. He has spoken about smart cards and biometrics at such venues as the Latin American Technology Conference at the World Trade Center in Mexico City, Mexico; CardTech/SecurTech and Electronic Funds Transfer Association (EFTA)/EBT Technology Conferences; Oracle Outer World; and the Biometric Consortium Conference.

1

Identity

Our names are at the core of our identities. But most of us did not choose our names; that was done for us at birth. One could argue that our names say more about our parents and their cultural milieu than they say about us. However, over time, we internally associate ourselves with our given names. Of course, names can change, but does a new name signify a change in identity? After all, names have power; the fabled trickster Rumplestiltskin knew that.* To know one's name is to have power over that person. Naming enables us to provide order to our world. Names convey meaning, and an inappropriate name conveys the wrong meaning. Names are labels used to categorize things and people: some categorizations are flattering and some are cruel. Names also provide a defining lens, and knowing a person's name helps us to associate that person with an identity.

How do we define ourselves? Do we use our nationality, our religion, or our profession? Do we ascribe to ourselves identities based on our homes or our possessions? Is it our core set of beliefs that defines us? Or is it the choices we make and the life paths we select? Of course, in everyday life we use identity proxies regarding both how we identify others and how they identify us. We use geography to refer to Asians or Europeans. We use language in referring to Chinese or English speakers. We often use religion, ethnicity, or membership to some organization, enterprise, or club. Our identities, and those of others, become associated with some general group, whether we agree with that association or not.

---

* Rumplestiltskin was a character in children's fairy tale who was promised a firstborn child if his name was not guessed.

Identity is a valuable commodity and it is precious to us. The concept of identity is much broader in scope than the content of the name. Naming conventions, such as surnames and personal names or nicknames, provide some identity structure, but they do not reveal the essence of one's identity. The true context and meaning of an identity can only be determined through reference to other factors because people exist simultaneously and often dependently in multiple social, political, cultural, and economic roles. Therefore, identity is complex and multifaceted, and its uniqueness must be viewed in comparison to the identity of everyone else.

Identity is often represented by an assortment of data that describes one's uniqueness. Therefore, we often are required to present documents that attest to our identities. Birth certificates record our entry into society; driver's licenses attest to our ability to properly drive an automobile; and passports confirm our citizenry. Each document conveys one's claimed identity for a specific purpose. Collectively, they serve as reference points that provide an acceptable level of certification regarding our identities.

It is increasingly important that we validate our identities when confronted by legitimate inquiries regarding who we are. Today, we cannot travel across geographical borders, rent a vehicle, check into a hotel room, buy a commercial ticket on an airline, or consult a physician without submitting at least one acceptable physical token that confirms our identity, whether that token is a passport, driver's license, credit card, or insurance card. In a world in which everyday activities are so dependent on providing identity credentials, it is critical that we vigorously protect that identity to keep it from being compromised, stolen, or misused by others.

In trying to prove our identity, we often rely on a set of breeder documents* such as a birth certificate. Millions of birth certificates are issued in the United States each year; yet most are issued in an insecure manner. In other words, birth certificates are public documents given with minimal controls to almost anyone who asks for them. A country as advanced as the United States cannot validate many of its birth certificates with any true certainty, nor can a government official confirm with certainty that the person claiming an identity on the birth certificate is who he says he is—at least, not without the help of other identifiers. A birth certificate simply does not represent a secured identity.† Unfortunately, the identity

* Breeder documents are so named because they are a source of identity that breeds other forms of identity credentials. In the United States, a birth certificate is needed to obtain a social security card. A birth certificate and social security card and one other picture identification (for example, a driver's license) enables one to obtain a passport.

† A secured identity is the verifiable and exclusive right to specific identity information.

vetting* process that precedes the issuance of critical credentials is often based on documents that themselves are not very secure.

In any discussion of identity it is important to distinguish between two separate identity concepts: identification and identity verification (also known as authentication). Identification is a process of identifying something or someone from a pool of many. It is a 1:n† application whereby one determines the identity of an individual by matching a query template of unique characteristics against all the templates stored in a given database. An identification process attempts to answer the question "Who are you?" Dependent on how tightly one has set the selection criteria, the accuracy of the database records (e.g., templates), and the size of the database (e.g., population) used in the process, it is possible, indeed probable, that an identification effort would yield numerous possible matches. Identification is regarded as a closed set if an unidentified person is known to exist in the database, or it is called an open set if an unidentified person is not guaranteed to exist in the database. The Automated Fingerprint Identification System (AFIS) is probably the best known identification system in the world. It has been in use in the criminal justice community since the first half of the 20th century. It is a minutiae-based fingerprint system that compares the measurements of a variety of features (such as ridge endings and bifurcations) in a given set of fingerprints against the stored minutiae templates‡ in its large database.

Unlike identification, verification seeks a one-to-one (1:1) match whereby an individual claims an identity and that claim is authenticated. Verification answers the question "Are you who you claim to be?" There are many ways to do this including the use of a badge or ID card, the use of a PIN§ or password,¶ an authenticated photograph, or via the use

* According to the Smart Card Alliance (SCA), there are more than 300 valid forms of government-issued IDs in the United States, making it difficult to distinguish between genuine and fake identification. See SCA's March 2004 report, Secure Identification Systems: Building a Chain of Trust, Princeton Junction, NJ.

† 1:n matching refers to the comparison of one biometric sample template with n reference templates, whereby n is a positive integer. If n = 1, then it is equivalent to a verification process; if n > 1, it is regarded as an identification process.

‡ A biometric template is the output of a feature extraction, usually expressed as a formatted, digital representation of an individual's unique characteristic.

§ PIN is a personal identification number, a secret numeric password used to authenticate a user to a given system.

¶ SCA studies have concluded that "90% of companies use passwords as the primary method of access control to information resources." See SCA's March 2004 report, Secure Identification Systems: Building a Chain of Trust (Princeton Junction, NJ: SCA).

of a biometric identifier. In biometrics, verification matches the sample template with the registered template, which corresponds to the claimed identity. The biometric algorithm used either accepts or rejects the identity claimed, or alternatively it could return a confidence measurement of the likelihood that the claim is valid. In the biometric world, identity verification is used much more frequently than identification applications. Generally, an identity verification search is significantly smaller than an identification search in template size, structure and scope, and in speed of results.

IDENTITY MANAGEMENT

In our globally connected world, business enterprises of all types are grappling with the challenges of managing secure access to proprietary information and critical applications scattered across a growing number of internal and external computing systems. Many are struggling to provide access to a wide range of users, including those outside the enterprise, without diminishing security or exposing sensitive information. The management of multiple versions of user identities across diverse applications makes the task even more daunting.

Identity management (or ID management; IDM) administratively verifies these user identities and controls their access to organizational resources across a variety of networked systems by associating user rights and restrictions with the established identities. Identity management governs the identity life cycle of individuals during which one's identity is established or reestablished. An identity is described with one or more attributes, and equally important, an identity can be destroyed. The typical steps in an identity life cycle* include:

- *Vetting the identity information*—This means cross-checking all identity documents that an individual presents and verifying the documents' information with the original issuing authorities. Depending on the application, it might include checking identity information against criminal watch lists and matching existing biometric information with the individual's biometric data.

* This information was well presented in a Smart Card Alliance white paper, "The Top 10 Hot Identity Topics," published in February 2006.

- *Enrolling the identity and creating a record that has privileges associated with it*—After the individual's identity has been verified, a record of the identity data is made and is associated with the privileges granted. Any personal data that is recorded must be treated as the private information that it is.
- *Creating an identity document or credential*—The appropriate authority issues a secure identity credential to the individual.
- *Using the issued credential to access associated privileges*—A key consideration is the safeguarding of the private information on the credential when it is used.
- *Revoking of privileges and terminating an identity*—An identity record is destroyed and all associated privileges terminated at some point. The reason could be as benign as retirement or graduation, or as threatening as criminal activity.

The underlying purpose of identity management is to make information (processed data) available to those individuals who are authorized to access it and to deny access to those who are not authorized. Nearly all identification methods in use today are based on one or more tokens or identifiers. There are (1) knowledge tokens (i.e., something you know) such as a password or PIN, or relatively obscure personal information such as your mother's maiden name; (2) physical tokens (i.e., something you have) such as a badge, card, passport, driver's license, key, or other device; and (3) behavioral or physiological biometric identifiers (i.e., something you are). Each of these methods has its merits and limitations, and each may be the appropriate solution in a given set of circumstances. But each method by itself offers only a single security factor. Strong authentication* demands a combination of authentication factors such as knowledge (passwords or PINs), tokens (smart cards), or process (e.g., biometric verification) used to verify identity under security constraints.

Knowledge tokens such as passwords and PINs enable delegation and the transmission over secure networks. However, they can be forgotten or compromised, and they do not provide nonrepudiation.† That is, its use does not guarantee that its assigned user actually inputted the PIN or

* Strong authentication refers to two-factor or three-factor authentication to deliver a higher level of authentication assurance.

† Nonrepudiation is the assurance that a party in a transaction or similar event cannot refute or deny the validity of that event.

password. Even X.509[*] digital certificates[†] and PKI[‡] systems used to prove one's online identity are not foolproof. A card, badge, key, or other physical token that is presented for access does provide a proxy for an identity claim, but each of these can be lost, stolen, or forged. Of course, some card tokens can securely store passwords or PINs, combining two separate security factors. Their joint use significantly improves the likelihood that they are legitimate; yet even together they remain only proxies for true identity verification. The card could be stolen and the PIN could be written down and acquired by a fraudster. The reality is that token-based systems present an individual who possesses a secure token or has special knowledge; but they do not completely verify the identity of the user.

Biometrics is arguably the only technology that can bind a person to an authentication event.[§] Knowledge and physical tokens cannot do that. Moreover, the person to be verified must be physically present at the point of identity submission. A biometric template could also be stored on a smart card, access to which generally requires a PIN; and together, they would provide three-factor security. When strong three-factor security is used in a transaction, the risk of fraud significantly declines and assurance of legitimacy substantially increases. Figure 1.1 illustrates the relative power of three-factor security.

The presence of a biometric template and PIN on a card badge with a smart (integrated circuit) chip does not mean that every application or even every transaction would necessarily have three-factor security. For convenience or practicality, some applications might use only the biometric or use only the PIN with the card. For example, a financial institution might require a user to use only his biometric identifier for access to the bank's own ATMs, but it might require the user to use both his biometric identifier and his PIN when remotely accessing financial records such as with home banking.

Identity and access management have become more complex as digital identities take on the increasingly important role of specifying how

[*] X.509 is a digital standard, specifying certificate structure. The primary fields of the standard are ID, subject field, validity dates, public key, and certificate authority (CA) signature.

[†] A digital certificate is structured according to the X.509 guidelines.

[‡] The public key infrastructure provides for a digital certificate that can identify an individual or an organization, as well as provides directory services that can store certificates, and revoke them, as necessary.

[§] This assumes a noncompromised biometric system.

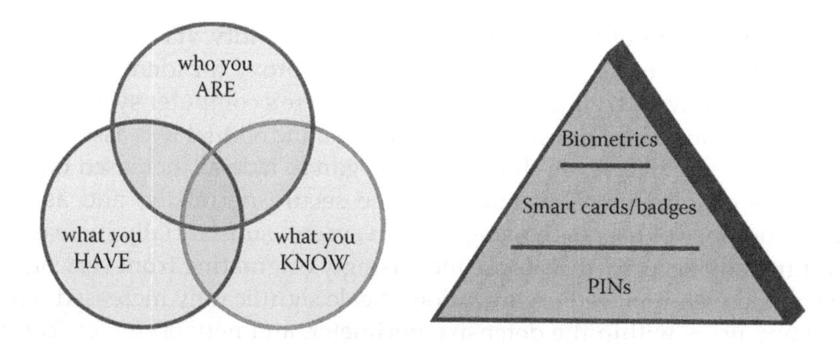

FIGURE 1.1 Relationship of three-factor security.

users interact with computer networks. Advances in the area of distributed computing, especially through the Internet, have resulted in a proliferation of applications and other mechanisms for accessing information in a typical enterprise. Naturally, most enterprises want to provide secure access to information assets for employees, corporate partners, and customers while positioning themselves for expansion, reducing access costs, strengthening security, and complying with regulatory requirements. Coping with these challenges can prove to be a daunting task. Various strategies have been derived to safeguard secure access.

To better understand the recent transition in information security strategies, it may be convenient to categorize them in three conceptual security strategies as presented by the security industry: (1) security of exclusion, (2) security of inclusion, and (3) security of accountability. Let us discuss each.

Security of Exclusion (SOE)

The concept of information security used to be about building security perimeters and fortifying them to thwart any attempted penetration. As the name implies, the security of exclusion concept is about denying access to unauthorized people in order to protect information. It represents data protection to ensure the confidentiality, integrity, and availability of that information. It is a defensive strategy focused on the security infrastructure, not unlike a *lockdown*. After all, information security used to be about safeguarding information within established secure perimeters, strengthening them with guards and walls, and defending against all unauthorized access attempts.

A key feature of SOE is that it infers one identity via an authorization. In the past one's physical location was a proxy for identification. If an individual were trying to access an enterprise's computer system from within the confines of the enterprise, then that individual was assumed to be an employee. However, that concept began to fade as increased online activities enabled more avenues into the secure perimeter, and as the actual number of users accessing a given system substantially increased. Additionally, security threats are increasingly originating from very organized and well-financed organizations. Scale significantly increased with so many users within the defensive perimeter, and networking concerns grew with many more avenues into the perimeter, rendering this mode of defense porous and significantly complicating the security system. So SOE has continued to lose favor as the key information security strategy.

SOE identifies potential risks and reacts to them. It establishes and maintains safeguards to ensure that breaches do not occur; and it suggests a tightly controlled infrastructure. Stout firewalls, antivirus or antispyware systems, intrusion-detection, and antispam programs are all basic tenets of a SOE strategy. It is very location oriented, defining authorization access protocols, not one's identity per se. This is not a strategy that engenders opening the security infrastructure for information collaboration. SOE protects the infrastructure, but does not necessarily safeguard the applications or the data. SOE still has its place as a supporting security practice, but it is no longer sufficient as a standalone strategy.

Security of Inclusion (SOI)

SOI is an identity-based model focused on providing the right level of access to legitimate users at the right time; it is sometimes referred to as security of enablement. SOI represents a shift from a lockdown mentality to one that confidently encourages online access. Most important, SOI requires verification of one's identity. Location is no longer an appropriate proxy for one's identity (e.g., an employee). From a practical standpoint, the Internet in general and private networks specifically have eliminated the concept of location as a significant security deterrent.

Thus, SOI is closely aligned with identity management, and SOI addresses both application-level and network-level identification. SOI enables enterprises, systems, and processes to open up to more users. Information actually tends to improve in terms of quality and organization when released to a large variety of users. Employees, customers, and suppliers each have legitimate needs to access enterprise information from time to time.

SOI is not an open free-for-all. SOI implies an infrastructure that controls processes and affords access only to legitimate users. It manages information flow of third parties. Moreover, it can provide access to only those resources for which an individual has access privileges. Those privileges can be revoked or temporarily blocked as roles change. Of course, the enterprise can track and audit the activities of a given identity to ensure compliance to its policies and procedures. Moreover, SOE and SOI are not mutually exclusive. Many enterprises integrate both strategies.

Security of Accountability (SOA)

Just as with SOI, security of accountability (SOA) is an identity-based model, but it is focused on compliance automation monitoring, controlling information flow, and reporting on its performance. SOA affixes accountability for assertions about an individual because those assertions will be relied upon by other parties. With SOA, identity becomes an enterprise construct in network security. SOA implies delivery of controlled processes and full enablement for role-based access to information. SOA wants to know who did what and when, and it enforces policies as applied to those transactions.

As information access has evolved and become interconnected, security solutions have had to evolve. Since networking is a decentralized process with multiple origins of data and multiple certifying authorities, identify verification must decentralize as well, which will support autonomy and mobility, and will require industry cooperation and interoperability standards. Additionally, it may necessitate the apportionment of authority. After all, SOA is a prerequisite to a federated identity management system.* Federation will be achieved only through open standards that enable interoperability and shared information to support identity portability. Addressing the challenges imposed by multiple organizations for cross-domain authentication is a new approach for identity management. Expectations are that identity federation will enable users of one organization to securely and relatively seamlessly access systems or data

* Federated identity refers to an authentication process that shares identity information of an individual across multiple information technology (IT) systems and across multiple organizations. Each participating domain accepts the individual's credentials.

of other organizations by using shared identity authentication information across multiple systems.

The speed of change within the security industry is expected to continue; meanwhile security concepts and strategies are still evolving. Each of the three strategies—SOE, SOI, and SOA—has its place. The industry must determine interoperability and integration boundaries and learn to cope with rapid change. What is now obvious is that security is transforming into identity-based solutions. Indeed, Phil Becker, editor of *Digital ID World,* is fond of stating that "security is now derived from identity." He cites identity as "the fundamental foundation upon which benefits such as security can be built."*

If the key to security is identity verification, which also facilitates legitimate travel and trade, then it is critically important to efficiently and accurately authenticate identities. However, there are many ways to effect that authentication. Increasingly, biometrics, which compares unique patterns for each individual, is being applied to provide accurate authentication. The essential objective of biometrics is protection. Biometrics protect against unsolicited and inappropriate access such as intrusion, theft, fraud, terrorism, injury, or loss.

ACCESS CONTROL

Protecting the network and ensuring the integrity of one's intellectual property rank high on the priority lists of most large organizations. In addition, stringent privacy laws in many countries have imposed new levels of confidentiality in such industry groups as health care providers, insurance carriers, and financial institutions. As a result, identity management has become a critical component to ensure physical and logical access control.

Enterprises often take a bifurcated approach to security: physical access and logical access security (also called information security). Access control is a method that enterprises use to enable or restrict rights of access to physical facilities or networks and information. Physical access control refers to access to such facilities as buildings, secure areas, data centers, and laboratories. Logical access control refers to controlling access to information such as business data, intellectual property residing on a

* Phil Becker, "The Evolving Models of Security," Digital ID World, October 17, 2006.

computer, or other sensitive information (including employee records and enterprise financial information).

The overwhelming majority of organizations have taken some steps to secure their physical facilities and control access to them. There are a variety of ways in which employees and contractors present their credentials for physical access in the workplace. Most enterprises still rely on antiquated card-based badges that require a radio-frequency identification (RFID) reader posted at the entrance to a building, floor, or room. These proximity badges merely attest that someone carrying an issued token passed by, but they cannot definitively identify an individual. And since many employees in a given organization "tailgate" into a building or floor, the system often does nothing more than ensure that at least one of the tailgaters waved an issued badge.

Moreover, as poorly implemented as physical access can be, logical access controls have lagged physical access at many enterprises. This is particularly perplexing in that there is greater potential for a breach in information security than in physical security, and the liability for such logical access breaches could be substantially worse than a breach in physical security. Logical access control is crucial to the success of an enterprise because the preponderance of an enterprise's assets is usually stored in electronic form. Therefore, logical access control should be given an equal level of focus commensurate to physical access control. An enterprise's information is a precious commodity, and it should be protected and controlled. Logical access control is concerned with how authorizations are structured.

Many organizations still rely on individual, user-based mechanisms such as a basic ID–password combination. User IDs and passwords are generally employed in the mode of a challenge and response. That is, a user who is trying to gain access to a given system is usually prompted to provide his or her log-in name and the password associated with that identity. The process is based solely on knowledge. As the number of users and applications increase, supporting such a system becomes time consuming, unwieldy, and expensive. Under the requirement to improve security, users are sometimes compelled to use and remember longer, "unguessable" passwords. However, users quickly become frustrated by the need to remember multiple difficult passwords that may frequently have to be changed. The sheer volume of requests for lost or forgotten passwords, new registrations, and account terminations bogs down customer service and help desk personnel. Past studies have indicated that heavy network users have more than 20 passwords to recall; approximately 80% of users select a common password and create derivatives, and nearly 30% record their passwords in a variety of easily retrievable ways. These studies have indicated that call center expenses have increased significantly in recent years; and this imposes significant costs, unnecessary resource use, and loss of productivity.

Logical access control has evolved into the authentication, authorization, and audit of a user for a session. Moreover, access control authentication devices have evolved from basic log-in ID and password to digital certificates, security tokens, smart cards, and biometrics. There are multiple approaches to managing user privileges including discretionary access control, mandatory access control, rules-based access control, user-based access control, and role-based access control. These methods overlap to some extent. There are also several other forms of access control, such as content-based and context-based control, which we will not discuss, but remain a fertile area for further research and development.

- *Discretionary Access Control* (DAC)—Based on the idea that the owner of data should determine who has access to it. Resource owners and administrators jointly control access. This model allows for much greater flexibility and drastically reduces the administrative burdens of security implementation.
- *Mandatory Access Control* (MAC)—In the military, users are classified as having access to levels of security such as top secret, secret, or confidential. Access is granted only if the user has a security clearance equal to or higher than the classification of the resource being requested. Mandatory access control emerged from the clearance requirements of the military. With MAC, only individuals with administrative privileges can manage access controls. MAC is considered to be a highly restrictive access control regime, inherently well-suited to the highest security environments. The access control system can also consider more compartmentalized access criteria based on a "need to know" or what operations a user can perform (e.g., read only, read/write, or execute).
- *Rules-Based Access Control* (RAC)—Associates access controls with specific system resources, such as files or printers. In such environments, administrators typically establish access rules for end users coming from a particular domain, host, network, or IP address. RAC systems may use a MAC or DAC scheme, depending on the management role of resource owners. Access control lists are a common rule-based access control mechanism. Many firewalls use RAC systems by checking its rules base to determine

if a requested connection is allowed; if so, it continues the connection through the network; if not, it closes the connection.

- *User-Based Access Control* (UBAC)—Identity based. However, most systems that use UBAC grant access permission to users who employ an identity proxy such as a PIN or password. Basically, a system administrator defines access permissions to individual users based on their perceived needs.
- *Role-Based Access Control* (RBAC)—Usually groups access rights by one's role,* and the use of resources is restricted to individuals authorized to assume the associated role. An individual's set of roles translates into a set of access privileges. This can ensure that a user is granted the level of access that is appropriate for each role and business function for which the individual is responsible.

Role-Based Access Control (RBAC)

Even if an enterprise has incorporated single sign-on capabilities,† typical user-based identity proxies such as the ID–password solution remains a very weak security mechanism. What is needed is a low-maintenance system that automates routine administration and controls access across the network so that data security is ensured. RBAC is one option that is being increasingly adopted by many enterprises.

Because RBAC reduces some of the complexities of security administration in large networked applications, it has become a key model for information security. In some cases, authorization may mirror the structure of the organization, whereas in others it may be based on the sensitivity level of various documents and the clearance level of the user accessing those documents. In many enterprises, the staff can be categorized into a relatively small number of predefined categories or groups, each with functional roles. Sometimes the groups and the role are synonymous; but other times the roles may be defined with fixed access permissions that any group or individual may assume for a period of time, in accordance with some predefined criteria. Oftentimes, the defined role matters more than the official position of the group member. Role-based privileges can be entered

* A role in this context is defined as a function or job assignment.

† Single sign-on (SSO) is a method of access control that enables a user to authenticate once and gain access to the resources of multiple software systems. The SSO process permits a user to enter one name and password in order to access multiple applications. The process authenticates the user for all the applications to which they have been given rights and eliminates further prompts when they switch applications during a particular session.

and updated quickly across multiple systems, platforms, applications, and venues—right from the security manager's desktop. By controlling users' access according to their roles and the attributes attached to those roles, the RBAC model provides a companywide control process for managing information assets while maintaining the desired level of security.

Computer systems can prescribe not only who may have access to a specific system resource, but also the type of access that is permitted. Users take on assigned roles such as physician, nurse, teller, or manager. The role of physician can include performing diagnoses or prescribing medications and treatments while the role of researcher might be limited to gathering anonymous clinical information for studies. The process of defining roles should be based on a thorough analysis of how an organization operates. In a medical office, for example, a receptionist is going to need access to different types of information and have different access rights than a physician. But in general, all receptionists need the same access rights.

The use of roles to control access can be an effective means for developing and enforcing enterprise-specific security policies, and for streamlining the security management process. User membership into roles can be revoked easily and new memberships established as job assignments dictate. Role associations can be established when new operations are instituted, and old operations can be deleted as organizational functions change and evolve. This simplifies the administration and management of privileges; roles can be updated without updating the privileges for every user on an individual basis.

AUTHORIZATION AND AUTHENTICATION

Once roles with access rights have been assigned, it is essential to authenticate identities. Based on the authentication of a user, access rights will be applied. By authenticating one's identity and the associated role description of an individual, we can determine what privileges he is authorized to access. Authorization is the process of determining access rights based on the authentication of a user; it permits or denies access to an object by a subject. Authentication establishes the validity of the identity given. It is based on identification and other security information, such as a password, which can be provided by the user, system, application, or process.

Access control can be enforced based on authentication of the individual and authorization of access rights. There must be an acceptable level of trust established that a person is who he claims to be. For an identity management system to be effective, trust must be established in an organization's online environment, especially in identities. To establish this trust, the unique attributes or credentials must be bound to a unique identity. This binding must be proven by authentication, which is achieved by verifying the identity of a user. Authentication techniques range from basic schemes employing user ID and password combinations to strong authentication schemes like two-factor authentication, digital certificates, smart cards, and biometrics.

The goal of authentication is to provide "reasonable assurance" that anyone who attempts to access a physical facility, system, or network is a legitimate user. In other words, authentication is designed to limit the possibility that an unauthorized user can gain access by impersonating an authorized user. Highly sensitive or valuable information demands stronger authentication technologies than does less sensitive or less valuable information. In a federated identity model, trust becomes even more crucial because companies need to manage and distribute identity information with one another in a controlled manner. This trust is achieved via authentication. Trust management* and federated identity management provide a standardized mechanism for simplifying identity management across company boundaries.

Because of their security, speed, efficiency, and convenience, biometric authentication systems have the potential to become the new standard for access control. Biometrics replaces or supplements knowledge and possession authentication with a person's physical or behavioral characteristics. Biometrics can be used in any situation where identity badges, PINs/passwords, or keys are needed. Biometrics offers some clear advantages over traditional identity methods:

- Biometric traits cannot be lost, stolen, or borrowed.
- Generally, physical human characteristics are much more difficult to forge than security codes, passwords, badges, or even some encryption keys.
- Biometrics guard against user denial—the principle of nonrepudiation—by providing definitive recognition of an individual.
- Biometrics cannot be delegated or shared. Its use proves that the individual in question was present for a given transaction.

* Trust management is a term that appears to have been coined in an AT&T research paper, "Decentralized Trust Management," in 1995 by Matt Blaze, Joan Feigenbaum, and Jack Lacy. It is the shared belief in the competence of an entity to act dependably, securely, and reliably within a specific context on behalf of another party.

- Identity verification can eliminate the need to carry a token or remember a password, although all three can be used.
- It is highly resistant to brute force attacks.
- Biometrics is the only technique available today that can determine if a person is who he *denies* he is or if he has pre-enrolled. This is the concept of negative identification.*

Biometric technologies offer enhanced security, convenience, and ease of use over traditionally used identity tools. Biometric authentication is becoming a security mainstay. The growth of the nascent biometrics industry and the widespread adoption of biometric technologies by the government and commercial enterprises are proceeding at an accelerating pace. Biometric technology may have begun its life cycle with governments and law enforcement applications, but it has become truly mainstream. With the growing demand for enhanced security combined with improved accuracy of biometric systems and declining prices, biometric solutions are becoming more commonplace. Biometrics is not just about security, it is about convenient security. Consider the following:

- Biometrics is now used in a rapidly growing number of countries to secure borders and to authenticate travelers and transportation workers. All visitors to the United States and to Japan now obtain tamper-resistant visas with biometric identifiers.
- Over 90% of all ATMs in Japan use biometrics (i.e., vein pattern recognition) to authenticate their cardholders before allowing an ATM transaction.
- The U.S. government has incorporated biometrics into its smart access card to be carried by the U.S. Armed Forces and other federal government employees. As part of the response to HSPD-12,† NIST‡ issued FIPS 210,§ which incorporates the special publication *Biometric Data Specification for Personal Identity Verification.*

---

* Negative identification provides evidence establishing that an individual is not already known to a given system or is located in a specific database.

† Homeland Security Presidential Directive 12 required all government employees and contractors to be issued credentials that would be interoperable across the U.S. federal government.

‡ NIST is the U.S.-based National Institute of Standards and Technology.

§ FIPS 210 defines the personal identity verification (PIV) card applications. PIV cards are secure smart cards used primarily for identity applications; the PIV standards were developed by the U.S. federal government, but now, they are used across a variety of countries and enterprises.

- The State of Queensland in Australia is using facial recognition biometrics on its new digital driver's licenses to improve security and hopefully reduce identity fraud for over 3 million licensed drivers.
- Nigeria re-registered the nation's 60 million voters linked with biometric identifiers to prevent fraudulent multiple registrations; a similar concept was recently implemented for the 80 million citizens of Bangladesh.
- Ticket holders at Orlando's Disney World now use a biometric reader at the turnstiles upon entering the park to ensure that the users of their multiday passes are the original purchasers.
- Employees of United Airlines use biometrics for their employee time and attendance applications.
- Call centers around the United States are using voice recognition to help them increase security vis-à-vis the sensitivity of the call transaction being handled.
- The Federal Railroad Administration's (FRA) Office of Research and Development has developed a biometric-based Locomotive Security System (LSS) to evaluate as a mechanism to prevent unauthorized use of locomotives.
- At the U.S. Armed Forces's Baltimore Military Entrance Processing Station, enlistees sign their enlistment contracts biometrically, as the military transitions to paperless recordkeeping.
- Kroger Company, based in Cincinnati, Ohio, has been using biometrics for years at selected sites. At some of its stores, such as in College Station, Texas, repeat customers can pay with their biometric identifiers, and in other locations customers can use their identifiers to cash payroll checks.

With continued enterprise decentralization, growing e-commerce applications,* and an increasingly mobile society, there will be greater demand for identity verification than ever before. For many applications, biometrics may be the only existing technology that can address the level of authentication that will be required. The security solution must start with accurate vetting of an individual's identity and follow that with appropriate identity verification processes. Nevertheless, we must not forget that security is a holistic concept and that biometrics is but one

* E-commerce is the buying and selling of goods and services on the Internet, especially the World Wide Web. In practice, this term and a newer term, e-business, are often used interchangeably.

component of the total security solution. Further, biometrics as a tool is not without its drawbacks, including additional costs, the requirement for specialized hardware, concerns over privacy and usability, and fear of system compromise. As with all technologies, there are ways to manage costs and ameliorate its shortcomings. In the succeeding chapters, we will discuss the potential of biometrics, but we will do so with a full understanding of its limitations.

2

Biometric Modalities

Biometrics means "life measurement." Biometrics refers to automated methods of recognizing individuals based on measurement of their physical or behavioral characteristics.

A biometric system is basically an automated pattern recognition system that either makes an identification or verifies an identity by establishing the probability that a specific physiological or behavioral characteristic is valid. It is based on the use of unique and measurable physiological or behavioral characteristics. Physiological characteristics include, but are not limited to, a person's vein patterns, facial structures, ocular characteristics, hand geometry, or fingerprint. Behavioral characteristics are based on an individual's unique actions captured over a period of time including such traits as signature dynamics, voice, gait, and keyboarding dynamics.

There are many real-world applications where security is a strong requirement, and reliable identity authentication is critical to that security. Token-based methods, including badges or passwords and personal identification numbers (PINs), tend to rely on surrogate representations of personal identities. Biometrics is considered a more natural and reliable solution for identity verification situations. Therefore, a biometric component for identity verification has become a critical enhancement for many security systems.

This chapter will focus on biometric requirements and then will present the reader with a survey of biometric modalities.* The survey will include both traditional physiological and behavioral biometric

* A biometric modality is a type or class of biometric system such as vein pattern recognition, fingerprinting, or facial recognition.

modalities, as well as some nontraditional biometric solutions. Finally, the chapter will apply the general requirements for a good biometric to each of these modalities in a comparison chart.

WHAT MAKES A GOOD BIOMETRIC?

There is no single biometric modality that is best. The appropriate biometric type for a given application depends on many factors including the biometric application (identification or identity verification), security risks, population of users, and user circumstances. Moreover, the biometric modalities are in varying stages of maturity in terms of implementation experiences. So a key question might be "What makes a good biometric for a given application?" Obviously, that would depend on the evaluation criteria one uses. There are seven evaluation criteria that help define a good biometric modality: (1) uniqueness, (2) permanence, (3) universality, (4) collectability, (5) acceptability, (6) performance, and (7) resistance to circumvention.

Uniqueness is the degree to which a biometric identifier differentiates one individual from another. Each individual variation of the biometric attribute must be unique to some extent. That requires a wide difference in individual patterns among the population. There is a big difference between categorization attributes (sometimes called soft biometrics)[*] and biometric attributes. Height, weight, eye color, and hair color are differentiating attributes for categorization; but they do not present adequate points of distinction to be useful for anything more than a broad categorization. Achieving an appropriate level of attribute distinctiveness for biometric data requires an array of statistically independent features, as measured by degrees of freedom.[†]

It should be pointed out that most scientists believe that all humans are unique in terms of distinctive features such as their vein patterns, fingerprints, retinas, and irises. However, there are billions of people on this planet and yet the largest single database in existence is the Federal Bureau of Investigation's (FBI's) fingerprint database for approximately 60 million individuals. Additionally, biometric templates can only store

[*] Soft biometrics are methods that identify individuals based on visual traits that assist in classification within a database but are not sufficiently distinct to be a unique identifier. This includes such traits as gender, height, eye color, and ethnic group.

[†] The term *degrees of freedom* refers to a measure of the number of independent variables on which the precision of a parameter estimate is based.

a limited number of distinguishing features of a given individual, and it is inevitable that certain unique properties acquired by a biometric sensor will be omitted, thus reducing the uniqueness of that individual. Moreover, every biometric identifier has some theoretical upper limit in terms of its distinctiveness. The uniqueness of any biometric characteristic may be influenced by the variability of the characteristic, the data collection used, and the passage of time.

Permanence refers to stability of the biometrics' attributes over time. How well does a given biometric attribute resist changes due to aging, injury, disease, and other factors? If a given attribute is less resistant to change, it would have to be periodically updated. Some biometric identifiers remain generally constant over one's life, once the individual reaches a basic maturity (beyond childhood), limiting the need for re-enrollments. On the other hand, aging significantly impacts some biometric modalities (e.g., facial recognition), albeit that is a slow process. But it must be stated that any human characteristic can change for internal or external reasons and that no biometric modality can guarantee permanence.

Universality describes how commonly a biometric is found in an individual and how readily it can be used. A good biometric attribute is one that is found in all human beings, and its usability does not vary significantly. If every person in a given population is able to present a specific biometric identifier for recognition purposes, then that biometric modality would be considered truly universal. However, no biometric modality is totally universal, although some identifiers are more universal than others. The biometric characteristic should present a variety of independent features. There is always a group of individuals who do not possess a specific biometric attribute; for example, finger biometrics does not work for individuals with missing hands. Thus, there is some percentage of the population that simply cannot be enrolled for a given biometric modality. Moreover, some biometric identifiers deteriorate with age, disease, or other physical changes.

Collectability indicates the practicality and ease of acquiring a biometric trait for quantitative measurement. The attribute should be suitable for capture and measurement, and must be convenient for the individual to present to the biometric sensor. Sometimes, an individual cannot provide a quality sample at enrollment, and sometimes he is unable to submit a biometric identifier that can be successfully matched to his enrollment template. The most common biometric modality today is the fingerprint; however, 2% to 5% of given population do not possess legible fingerprints for purposes of collection.* Some people just have fingerprints too faint to reliably measure for either verification or identification applications. On the other hand, vein patterns throughout the human body have proven relatively easy to acquire for pattern measurement (with the consent of the individual, of course). However, only vein patterns located in an accessible area such as one's hands or wrists are used in commercial applications today, as attempting to measure vein patterns elsewhere has proven inconvenient or impractical.

Acceptability indicates the degree of public acceptance and approval for a given biometric modality. That is, does the user population accept sample collection routines during enrollment and during verification or identification transactions? Generally, nonintrusive biometric techniques tend to garner greater levels of user acceptance.† This is a very important criterion because user acceptance is critical to the success of any biometric implementation. Another constraint to acceptability is cultural context. For example, some cultures attach a stigma to fingerprinting or they feel that placing their hand on the same hand geometry readers that dozens of other people are using is unhygienic.

Performance indicates the accuracy, speed, and general robustness of the system capturing the biometric technique in varied environmental circumstances. Performance is an umbrella term referring to various measurements. It is generally true that speed reduces accuracy, since the attainment of one of these measurements is somewhat inversely proportionate to the other. Accuracy of biometric systems is usually defined by their false accept and false reject rates. Accuracy is affected in the data collection process by environmental (e.g., lighting, shadows, background noise) and individual (e.g., positioning, cooperation) causes, as well as the reliability of the biometric sensor used and the distinctiveness of the features extracted. It is also affected by the lack of invariant representation of a person's biometric identifier.

Speed is affected by the size of the database for matching, the computational power of the biometric system used, and any techniques used

* This interval is used here because studies widely vary as to the actual percentage, depending on the targeted group. The key point is that among the traditional biometric modalities, fingerprint systems have some of the highest uncollectible rates.

† Intrusive refers to something unwelcome or forced; nonintrusive suggests that the biometric technique is acceptable to most people. The term *invasive* is often confused with intrusive, even among biometric solution providers. The use of invasive is basically a malapropism, as invasive means "penetrating into the body," and no biometric identifier in use today is truly invasive.

to increase speed (e.g., binning and filtering). However, while increasing speed, the use of these techniques might introduce new errors. A discussion of speed is usually confined to the biometric system's data processing speed and how quickly it will return an accept or reject decision. That would include the processing speed of a reference template matched against a sample template. However, a true measurement regarding speed should be user throughput of a biometric system, and that would include the entire effort of an individual using a biometric system from inputting a card or PIN to inserting or aligning a finger or hand, processing it, and obtaining a decision.

Resistance to circumvention refers to how hard it is to spoof or otherwise defeat a biometric system. Biometric systems can be fooled, albeit it is getting harder and harder to do so. Some biometric systems have been focused primarily on user convenience and offering low-cost deployments. This has been particularly true of some low-end fingerprint systems. However, low-cost, plug-and-play systems are usually the ones that are easiest to circumvent. Again, we must recognize this reality and take steps to minimize how this can occur. Circumvention attempts include artificially created identifiers, attacking via an input port, or attacking the database. Protection includes introducing liveness detection, encryption, and challenge–response mechanisms, among other solutions.

These aforementioned seven criteria can provide an unbiased basis for evaluation of a given biometric modality. A more pragmatic view of what makes a good biometric might relate to application requirements and constraints. Each biometric modality tends to address applications and requirements uniquely and therefore differently than other modalities. For example, facial recognition works well at ports of entry for surveillance applications. For such an application, there is no requirement for user cooperation or any user involvement. A similar situation exists with voice verification for an application activated over a telephone. In the next section of this chapter, we will review the traditional and many nontraditional biometric modalities known today.

PRIMARY BIOMETRIC TYPES

No one has yet devised a perfect biometric system. Various types of biometric systems are being used for real-time identification and identity verification applications. There are a variety of biometric techniques, each with its own set of advantages and disadvantages. Thus, it is the specific application for which the biometric technique is intended that should determine the most appropriate biometric system. Further, the uniqueness of each biometric type means that a biometric should be evaluated individually relative to the application.

Why are some biometric modalities considered primary or traditional while others are generally listed as emerging or nontraditional modalities? Different authors categorize biometric identifiers in different ways. The criteria that I have used are commercial availability and a critical mass of users. If a biometric modality (for example, ear shape or gait recognition) has few or no commercial systems available for purchase, then I have placed it in the nontraditional, emerging category. On the other hand, vein pattern recognition has more than 100 million users globally, and there are multiple commercially available and viable vein pattern biometric systems. Clearly, vein pattern recognition systems should be categorized along with other primary biometric modalities.

Physical Characteristics

The primary biometrics based on physical characteristics currently include ocular recognition (that is, retina* and iris), facial recognition, fingerprints/palm prints, hand geometry, and vein pattern recognition.

Ocular Recognition

One's eyeball has unique and fairly accessible identifying characteristics that remain constant over time. Ocular or eye recognition focuses primarily on the retina (the innermost layer of the wall of the eyeball), as shown in Figure 2.1, and the iris (the colored disc surrounding the pupil in the front of the eyeball), as shown in Figure 2.2. Eye recognition systems are among the most accurate biometric techniques available, and virtually anyone with a healthy eye can be readily enrolled. No iris or retina pattern is the same, even between the right and left eye, or between identical twins. The public generally perceives these two biometric methods to be similar although they are dramatically different in operation and in the level of user cooperation needed.

Retinal biometrics, also called retinography, verify one's identity by reviewing the unique pattern of blood vessels on the retina. Retina scans

* There are few, if any commercial systems today that support retina recognition; but there are some users with older systems. However, I am including this modality in the traditional list because of retina recognition's history.

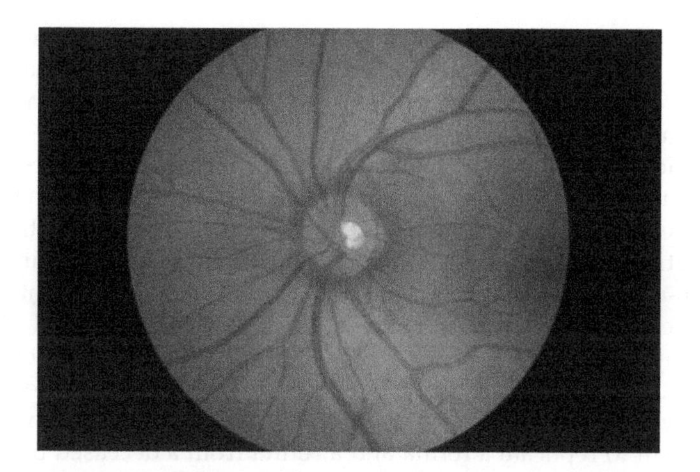

FIGURE 2.1 Photo of a retina.

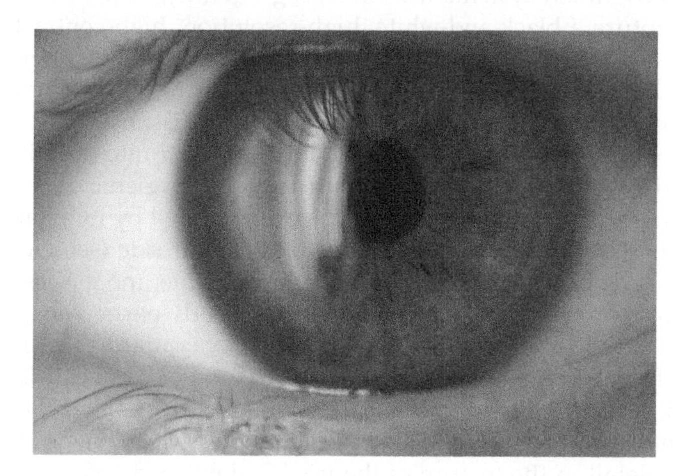

FIGURE 2.2 Photo of an iris.

capture the blood vessel patterns at the center of the retina, referred to as the optic disc.* These blood vessels tend to be stable during one's lifetime. The retina scanner shines a low-intensity beam of infrared light into the eyeball. A user must remove his glasses, keep his head stationary with his

* The optic disc is the small, circular, optically insensitive region in the retina formed by the meeting of the retinal ganglion cell axons (nerve fibers) as they enter the optic nerve.

eye close to the camera (about 30 cm), and he must focus on a specific target while the infrared beam is aimed at his pupil (similar to what occurs at an eye exam). A camera measures the reflected light and it can capture approximately 100 data points.

In addition to being highly accurate, retina scanning produces a very compact template. Thus, this modality's template requires only a fraction of the memory storage that most other modalities require. On the other hand, retinas are susceptible to such diseases as glaucoma, retinopathy, and age-related macular degeneration (AMD), which occurs at a high incidence among the elderly. Even though there is no physical discomfort, many users are not comfortable with retina scans, as it is seen to be somewhat intrusive and cumbersome. Retina scanning has been used primarily in high-end security applications, since there is no current mechanism to replicate a retina and a retina from a deceased body cannot be used.

Iris recognition systems use small, high-quality, near-infrared* cameras to capture a black-and-white, high-resolution, high-contrast photograph of the iris. The near-infrared illumination reveals patterns even for dark irises with no discernible patterns during visible light. Once the image is captured, the iris's elastic connective tissue is analyzed, and the distinctive features are extracted and translated into a digital form. Iris-based systems are relatively nonintrusive and hygienic.

The iris is the annular region of the eye bounded by pupil at its center and by the sclera (white of the eye) on its outer side (see Figure 2.2). Every iris is distinct. An iris is highly stable over time, and it is less subject to wear or injury than most other measurable body parts. The iris offers a data-rich structure composed of fibrous and vascular tissue including freckles, furrows, pits, rings, corona, and striations; however, iris color is not used. It is the measurement of these characteristics and their intra-spatial relationships that provide useful data for verification or identification. The detail in the texture of the iris is established before birth, and it appears to be caused by random processes, referred to a chaotic morphogenesis.† The human iris begins to form during the third month of

* Throughout this book the term *near-infrared* is used. There is no hard definition that is universally accepted; however, for vein pattern use, light that falls into the wavelength of approximately 700 to 1200 nm is usually classified as near-infrared light.

† The term *chaotic* refers to the dependence on the conditions in embryonic genetic expression. Morphogenesis is the process of cell development into different tissues, organs, or structures.

gestation, and the structure of an iris is completely formed by the eighth month after birth.

Whereas most biometrics depend on 13 to 60 distinct characteristics, the complex iris is said to have 266 degrees of freedom of textural variation, rendering it potentially the most accurate biometric modality in use today. Indeed, a person's iris contains approximately six times as many unique, measurable characteristics as fingerprints.* In reality, iris-based biometric systems do not yet approach this level of accuracy. Iris scanners can measure the characteristics in the iris at a distance of a meter or more. The muscles in the iris control the amount of light entering the pupil, and thus regulate the size of the pupil. The iris reacts to the stimulus of light with involuntary reflexes throughout the iris muscles. Some iris systems use this in a challenge–response effort to guarantee a living eye and not an artificial image.

The collection of an iris image does require more training and attentiveness than most biometric modalities. Although the technology can be highly accurate, poor quality images caused by drooping eyelids, enlarged pupils, off-centered irises, or poor resolution and blurriness are not uncommon. These difficulties can result in higher than normal failure to enroll rates in some systems. Most iris recognition systems are somewhat confining based on the required read distance,† the need for special lighting, and the desired camera angle (position of the eye in relation to the camera). Also, iris scanning systems tend to be among the most expensive biometrics. Nevertheless, this biometric can work well in either identification or verification modes. It is less intrusive than retina scanning and another key advantage is that it can occur passively, that is, without active effort by the subject. Current systems can be used even in the presence of conventional eyeglasses and contact lenses; however, dirty glasses or severely scratched lenses can interfere with the recognition process.

Iris scanning was designed for physical access control applications, but it is now used in a variety of venues including high-security facilities, health care, and commercial banking. Indeed, iris recognition is being used in a growing number of access applications in both government and commercial arenas. Figure 2.3 depicts an authentication solution in use in Afghanistan.

---

* William J. Krouse and Raphael F. Perl, Terrorism: Automated Lookout Systems and Border Security Options and Issues, Congressional Research Service (#RL30084), June 2001.

† Read distance is the distance from the biometric sensor to the subject's physical attribute.

FIGURE 2.3 Photo of a U.S. soldier and an Afghan man. (Courtesy of L-1 Identity Solutions.)

The recent lapse of iris patents should help to ignite additional competition, which may lead to a decrease in the current high prices for iris scanning systems. Improvements in ease of use and system integration are expected as new products are brought to market.

Facial Recognition

Commercial applications for automated face recognition technology first surfaced in the mid-1990s. The technology has made rapid advancements in the past decade. Facial recognition is a natural means of identity recognition among humans. Facial recognition systems attempt to find recognizable facial characteristics from photographs and video, reduce the key features to digital codes, and match them against known facial templates. Although various techniques can be used, facial recognition systems tend to emphasize facial regions that are less susceptible to alteration such as the areas surrounding the cheekbones, upper outlines of eye sockets, the distance between the eyes, and the sides of the mouth.

Facial recognition systems can process a two-dimensional (2D) or three-dimensional (3D) camera image, depending on the system. Face scanning can be accomplished at a distance for clandestine surveillance or in close proximity to the subject for forensic or overt identity verification. Two-dimensional facial recognition systems have not achieved high accuracy relative to some other biometric techniques, but they are constantly improving. In the past, such systems have been impacted by facial hair, glasses, lighting variances, significant weight gain or loss, camera angles, and face masks used for breathing. Three-dimensional facial recognition is not as dependent on lighting conditions or varied poses, including nonfrontal angles. While 3D systems are still developing, they have the potential for high accuracy, although they tend to be less nimble at processing data from large crowds. For 3D systems to acquire a larger commercial role, they must acquire more powerful and less expensive facial recognition sensors.

The advantages of facial recognition include high public acceptance of the modality, commonly available sensors (for example, cameras), not physically intrusive, and the ease with which humans can verify the results. A key advantage to facial recognition is that this biometric can be used anywhere that one uses a camera, from an automated teller machine (ATM) to high-traffic locations. An interesting by-product of this biometric is that the system administrator can retrieve a visual record of unauthorized attempts at entry. The disadvantages include its susceptibility to spoofing in large databases as the face can be obstructed; its sensitivity to lighting, facial expressions, and pose; its large templates (approximately 3,000 bytes or more); and the reality that faces do change with age. Moreover, facial recognition can threaten privacy as it can occur without the subject even knowing about it. Thus, it is somewhat controversial because it is the primary biometric modality used in covert surveillance in some very public and not-so-public places.

Facial recognition supports both identification and verification applications. For identification, an algorithm identifies an unknown face in an image by searching through an electronic mug book.* In verification applications, an algorithm confirms the claimed identity of a particular face. Proposed applications have the potential to impact many aspects of everyday life by controlling access to physical and information facilities, confirming identities for commercial transactions, and controlling the flow of citizens at ports of entry. It has been used successfully throughout gaming casinos since the late 1990s as a mechanism to identify banned gamblers. Facial recognition has been a well-used biometric identifier for government surveillance. As depicted in Figure 2.4, facial recognition continues

* An electronic mug book is an electronic file of identifiers or reference templates against which a sample template is compared on a 1:n basis to try to establish the identity of a previously unidentified person.

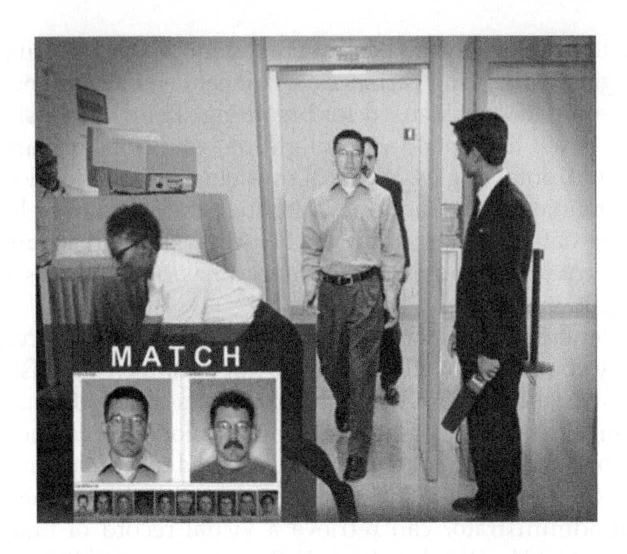

FIGURE 2.4 Access application for facial recognition. (Courtesy of L-1 Identity Solutions.)

to broaden its adoption to a variety of physical access and accountability applications.

Fingerprints

A fingerprint biometric is a digital version of the ink-and-cardboard method used throughout most of the 20th century. Thus, a fingerprint-based identifier is the oldest biometric method in use today; and due to its significant head start, it is the most widely used biometric modality. Fingerprint scanning includes the acquisition of a person's fingerprint characteristics and the quantification of the probability for authenticating the person's identity. Fingerprints are based on physical dermal structures defined before birth. The pattern of furrows and ridges on the surface of each finger (see Figure 2.5), minutiae points (e.g., local ridge characteristics that occur at a ridge bifurcation or a ridge ending), and image density are unique to each individual. The print pattern of a finger's top joint (the distal interphalangeal joint) is the primary focus. Historically, fingerprinting has been used for identifying criminals and for other forensic activities. Scotland Yard's Central Fingerprinting Bureau was established in 1901, and the FBI implemented the Automated Fingerprint Identification Systems

FIGURE 2.5 Sample fingerprint.

(AFIS) in the 1920s. Today, IAFIS[*] is used worldwide for government and commercial applications.

Because fingerprints are the most mature biometric technology, there are many vendors and product choices. Fingerprint systems are also known as finger scanning or finger imaging systems. Fingerprint-based systems work by scanning the tips of one or more fingers and comparing the scans of the finger against known images—the preestablished fingerprint template created during program enrollment.

Non-AFIS fingerprint recognition does not support the level of detail that AFIS does, and non-AFIS fingerprint systems are generally not appropriate for forensic activity. Traditional fingerprint sensing systems use a variety of techniques to capture a fingerprint image including capacitive,[†]

[*] Integrated Automated Fingerprint Identification System (IAFIS) can perform 100,000 comparisons per second, with a database of approximately 60 million individuals (with multiple templates per individual).

[†] Capacitive sensors form fingerprint images on the dermal layer of skin by using principles of capacitance (generating a small electric current). There are passive and active capacitance sensors.

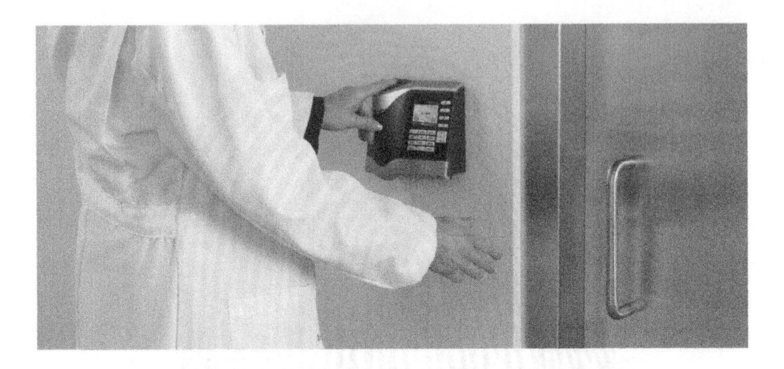

FIGURE 2.6 Health care worker gaining physical access. (Courtesy of L-1 Identity Solutions.)

ultrasound,* and optical† methods. A popular technique for direct-optical technology is the CMOS (complementary metal-oxide semiconductor)–based silicon sensor.‡

Fingerprint scanning has several advantages. Given its history and common use, most people understand the concept, and training is minimized. Figure 2.6 depicts a fingerprint system mounted on a wall adjacent to a secured entry point. The sensors that are used in fingerprint applications tend to be quite small, allowing some very useful form factors for fingerprint readers in laptop computers, mobile phones, and PDAs. Figure 2.7 is a photo of a highly mobile fingerprint solution. The fingerprint templates are also small, enabling a greater volume of template storage in limited memory situations, as well as enabling a very fast retrieval process. Most types of fingerprint scanning equipment have become relatively inexpensive. Moreover, as a biometric modality, fingerprint scanning can support verification and identification applications.

But fingerprint scanning has an extensive set of limitations that should be considered. Biometric industry authorities indicate that between 2% and 5% of the general population has some physical limitation in regard to using fingerprint imaging technology. Fingerprint ridges deteriorate with age and wear. Some occupational activities such as aggressively

* Ultrasound sensors use medical ultrasonography to create images via high frequency sound waves that penetrate the epidermal layer of skin.

† Optical sensors refer to the use of visible light to capture a fingerprint pattern image. This technique includes the frustrated total internal reflection (FTIR) method, which is the oldest and still the most used.

‡ CMOS or chip-based imaging is found primarily in low-cost fingerprint scanners generally intended for noncommercial-level verification applications.

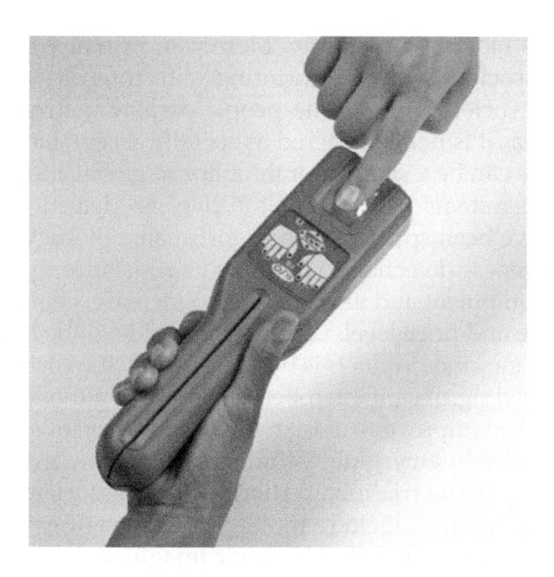

FIGURE 2.7 Portable Integrated Biometric Identification System (IBIS). (Courtesy of L-1 Identity Solutions.)

washing one's hands multiple times a day (e.g., surgeons) or constantly placing one's fingers in strong chemicals (e.g., hair beauticians) erode one's fingerprints over time.

Depending on an individual's environment, his fingertips may rather easily become dirty, oily, or cut. Depending on the type of sensor being used, dirt and oil could obscure the finger image, as could cuts, calluses, or scars on the fingertip. Dirty and oily fingers tend to render the fingerprint readers dirty and oily, and that might subsequently induce poor scans.* Fingerprint templates might be significantly influenced by dryness of the fingers or by the pressure or positioning of the fingers for a scan. Deteriorated fingerprint image quality may result in matching errors and overall system weakness.

It is well known that fingerprints have a latency property. That is, we leave our fingerprints behind on just about every surface we touch: from the drinking glass we hold at restaurants, to the doors we open, to the phone handles and keyboards we use at work. This can impact privacy and can encourage imposters to try to spoof the system, decreasing the

* In fairness to the fingerprint modality, a comprehensive cleaning effort systemically applied to the fingerprint plate will minimize much of this concern. This may or may not be an actionable solution, depending on the situation and application.

security of this biometric modality. Moreover, in many countries there is a cultural association of fingerprinting with forensics and the criminal element of society; thus, some people display a strong aversion to fingerprinting, as it is not considered as socially acceptable as some other modalities. This can be a significant inhibitor to global adoption.

Fingerprint systems have no built-in liveness detection, and numerous systems have been spoofed with unfortunate notoriety. If fingerprint systems add liveness detection to improve its resistance to spoofing, then the size of its equipment and its price tend to increase, neutralizing its key advantages: size and price level. Last, the downside to the high number of fingerprint vendors and products is that historically the various fingerprint scanners have had significant interoperability issues, although the situation has shown marked improvement with the introduction of new standards.

There are several emerging techniques that show great promise in alleviating many of the traditional shortcomings associated with fingerprints. These techniques include multispectral imaging (MSI),* 3D contactless† finger imaging, and 3D ultrasonic imaging.‡

Law enforcement has been determining subject identities for decades by matching key points of ridge endings and bifurcations. Personal computer (PC)–based fingerprint recognition devices are now widely available from many different vendors at relatively low costs. Fingerprint recognition dominates the network security market. Several states check fingerprints for new applicants to social services benefits to ensure recipients do not fraudulently obtain benefits under false names. New York State has over 900,000 people enrolled in such a system.

Very similar to fingerprint recognition, palm print recognition measures the prints in one's palm to verify identity. Palm print recognition has similar characteristics as fingerprint recognition, and it has a similar set of advantages and limitations. Because the palm is larger, more distinctive features can be acquired in comparison with fingerprints. On the other hand, palm print scanners tend to be bulkier than fingerprint

* MSI or multispectral illumination is an optical method that looks at the skin's surface and subsurface features, capturing raw images by employing illuminating lights of different wavelengths and different polarization conditions.

† 3D contactless is a finger imaging method that uses remote sensing to capture ridge-valley patterns. There is no physical contact between the finger and the sensor. Although the technology continues to advance, it remains somewhat vulnerable to high-quality spoofing (e.g., using a fake gelatin-based finger with someone's lifted fingerprint).

‡ The 3D ultrasound is based on matching paired images using internal fingerprint structures. It is difficult to spoof and has more tolerance to external conditions than most methods.

scanners and are usually more expensive. Many commercial fingerprint system providers have added palm print scanning to their biometric lineup of products. The FBI added palm prints to its IAFIS program.

As with fingerprint scanning, the three primary techniques for palm matching currently include minutiae-based matching,* correlation-based matching,† and ridge-based matching.‡ And as with fingerprint scanning there are a variety of sensor types—optical, capacitive, thermal, and ultrasound—that can acquire the digital image of a palm print.

Although some biometric practitioners might disagree, palm print imaging is effectively a subset of fingerprint imaging. The skin on one's finger and palm that is being scanned, and the technologies used to collect fingerprint and palm print data, and compare templates are nearly identical.

Hand Geometry

Biometric systems based on hand geometry are among the most commonly used systems. Hand geometry focuses on the physical structure (lengths, widths, thicknesses, and angles of the fingers and palm) of an outstretched hand. Typically, a user places his or her hand on a reader that is studded with guide pegs. The advantages of hand geometry are that it is intuitively easy to use, it typically uses 9 to 20 bytes of data for its templates (the smallest template of any biometric modality), its error rates are very low for false reject and failure to enroll, it has fewer privacy concerns than fingerprint or face recognition, and its readers work well in harsh environmental conditions such as commercial plants where users have very dirty hands. The modality is well suited for industrial locations such as warehousing or some manufacturing facilities (for physical access control and time and attendance applications) where its simplicity is a positive. Moreover, hand geometry can be readily combined with other biometrics, such as fingerprints and vein pattern recognition of the hand or fingers.

For better or worse, the human hand is not truly unique, causing hand geometry scanners to operate with a relatively high false accept rate (on the order of 0.1%). Additionally, its features vary over time (more than

* Minutiae-based matching is the most widely used. It relies on the location, direction, and orientation of each minutiae point.

† Correlation-based matching involves alignment of palm image templates and then determining the correlation of two palm vein images.

‡ Ridge-based matching reviews the geometric characteristics and spatial attributes of ridge pattern features in the palm's skin.

most biometric modalities). The human hand has 27 bones, and it provides adequate anatomical features for measurement to enable 1:1 identity verification, or 1:n identification within a limited database. The primary disadvantages of hand geometry include the relatively large size of the readers, its moderate accuracy rates, and the sanitation concerns of placing one's hands in the same place as dozens or even hundreds of other people that preceded the user. Since there is such a small amount of information measured, hand geometry is not appropriate for most identification applications, especially with large populations. Current research work involves identifying new features that would result in better discrimination between two different hands and designing a deformable model for the hand.

The basic procedure to use a hand geometry device is as follows: the user presents a credential (card or PIN), then places the palm of his hand on a flat surface that has guidance pegs on it. If the user's hand is properly aligned, the biometric reader can read the hand attributes. The reader then checks its database for verification of the user. The process (including credential presentation) usually takes 4 to 5 seconds.* Please refer to Figure 2.8 of the hand geometry scanner.

The relatively large size of hand geometry readers restricts their use in widespread applications such as those requiring a small user interface (e.g., home computer user). Hand geometry readers could be appropriate where users are perhaps less disciplined in their approach to the system, outdoors, or in harsh environments. However, its use with children can be challenging.

Hand geometry devices can function in extreme conditions, withstanding wide temperature variances, and are not impacted by dirty hands or dusty environments (as some fingerprint sensors can be). Modern hand geometry devices have been successfully manufactured since the early 1980s, placing hand geometry among the first biometric modalities to find widespread use. Over 300,000 hand geometry readers have been sold worldwide, including the San Francisco International Airport, U.S. nuclear power plants, military bases, universities, day care centers, welfare agencies, and hospitals. They are also used to streamline security and immigration procedures for over 90,000 users in the

* J. Holmes. L. Wright, and R. Maxwell, "A Performance Evaluation of Biometric Identification Devices," Sandia National Laboratories Report, SAND91-0276 (June 1991).

FIGURE 2.8 Hand geometry scanner. (Courtesy of Schlage.)

INSPASS* frequent international traveler system, 340,000 users at Ben Gurion International Airport,† and more than 50,000 dayworkers in an Israeli border control system.‡

Vein Pattern Recognition

Based on the unique patterns of veins in one's finger or hand, vein pattern recognition (VPR),§ also referred to as vascular pattern recognition,¶ provides the ease of use of hand geometry with much improved accuracy,

* The INS Passenger Accelerated Service System (INSPASS) card uses hand geometry for identity verification purposes. The INSPASS program facilitates airport congestion by speeding up the customs process.

† See www.eds.com/industries/homeland/downloads/idmanagement.pdf, accessed July 2009.

‡ See www.eds.com/services/identity/downloads/assuredid.pdf, accessed July 2009.

§ Some people refer to vein pattern recognition as vein geometry.

¶ Technically, retina scans are a form of vascular pattern recognition since it uses the pattern of blood vessels in the back of the eye to identify individuals; the same is true of facial thermography. For this reason, the term vein pattern recognition seems a more apt description.

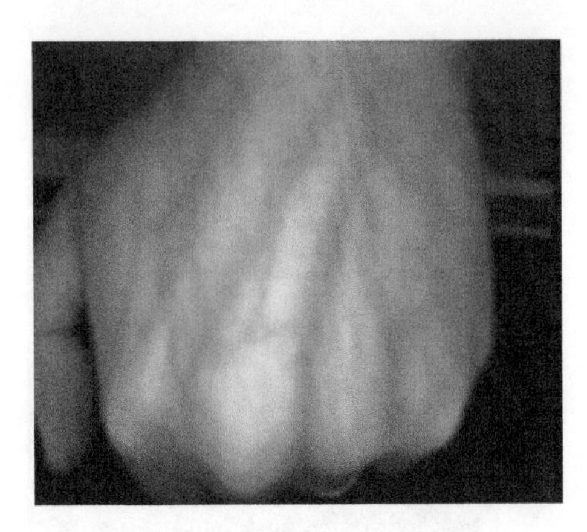

FIGURE 2.9 Vein pattern recognition. (Courtesy of Identica Corp.)

smaller readers, and a much more hygienic approach (see Figure 2.9). It uses near-infrared light generated from a bank of light-emitting diodes (LEDs) projected through an individual's skin to enable a high-contrast matching of vein patterns (e.g., vein branching points, branching angles, etc.) in one's hand or finger.

VPR systems scan the deoxygenated veins in one's palm, back of the hand, or fingers, extract key pattern features via contactless near-infrared optical sensor systems, digitize the extracted pattern recognition, and then match the transaction templates to the respective preestablished enrollment template. By measuring the veins under the skin with near-infrared spectroscopic imaging, it is very difficult for unauthorized persons to observe or capture this pattern, rendering VPR a highly secure method of identity verification.

VPR systems have won wide acceptance in banking, especially in Asia. Moreover, VPR systems have some very powerful advantages. First, there is no property of latency: vein patterns are located in the fingers or hands and cannot be seen by the naked eye. Vein sensors are highly accurate and durable. The sensors are looking below the skin, and they generally do not have issues with minor cuts and moisture. VPR systems reveal no significant performance degradation in harsh environments or when measuring sweating or mildly dirty hands.* VPR systems demonstrate very high accuracy rates, currently higher than some fingerprint

* Some "dirt" such as carbon from a copier machine or coal mine, or certain inks may obscure the infrared image.

imaging systems and nearly on par with iris recognition. Also, they are very difficult to spoof since blood needs to flow to register an image. Only a slight portion of the hand makes contact with the scanning device or its guideposts in order to align the finger or hand for consistent imaging. Since users do not touch the sensor surface, VPR systems are considered quite hygienic, especially in comparison to hand geometry or fingerprint systems. Some newer vein pattern scanners are totally contactless. VPR systems are extremely easy to use and require only a moderate level of training on the part of the user. There are no special user requirements for enrollment or ongoing use. Last, VPR systems have demonstrated exceptionally high usability rates exceeding 99.9% of most populations.* Usability refers to the percentage of a given population that is capable of using a biometric system. Given that some fingerprint systems have a usability rate slightly over 95%, a 99.9% rate is quite significant.

As with all biometrics, VPR systems have their limitations. Direct sunlight or other ambient bright light can impact VPR systems, which is why some VPR systems have been positioned primarily for indoor use. However, many VPR system providers have designed coverings for outside access systems, and some even have semitransparent coverings to limit ambient light indoors. Some VPR systems do not offer robust algorithms that support 1:n applications.

Behavioral Characteristics

The primary biometric modalities based on behavioral characteristics include dynamic signature verification, keystroke dynamics, and voice verification. Each is discussed in the following sections.

Dynamic Signature Verification

Most of us sign our names in a very recognizable and repeated fashion. Signature recognition is a behavioral biometric that takes advantage of that human tendency. Further, throughout most of the world, signatures are legally binding. However, since sight verification alone can be compromised, dynamic signature verification (DSV) measures both dynamic behavior such as speed, stroke order and direction, and pressure of writing. DSV differs markedly from paper-based signature recognition, which merely compares the visual aspects on a given signature. DSV can also

* Testing by the International Biometric Group (IBG) found the Hitachi finger vein and the Fujitsu palm vein both scored an availability rate of 99.92% based on enrollments.

record static behavior including shape and pattern of the signature, and "pen in air" movements. The captured values associated with one's handwritten signature enable online authentication of individuals. And it can be integrated into the approval process for a variety of procedures.

It is very difficult to spoof the behavior of signing one's signature. DSV is an accepted biometric in banking and financial applications, especially for document authorizations and electronic paymenting. Also, electronic signatures are supported by law in some countries, such as the United States. Thus, DSV has growing social and legal acceptance. It is not considered hard to use. It offers an attractive option as added security for e-business applications and other applications where one's signature is an accepted method of personal authentication. Indeed, thousands of sign pads, such as the one in Figure 2.10, have been installed at venues such as bank tellers, at point-of-sale (POS) terminals at retailers, and on the counters of telecommunications companies. Digitization of one's handwritten signature throughout the signing process is a natural means for paperless signatures, and it is continues to grow in acceptance and use.

As with all other biometrics, DSV has some limitations. One's emotional state, fatigue, or illness can affect its accuracy, as individuals do not provide a consistent signature over time. Moreover, there is growing

FIGURE 2.10 Dynamic signature verification. (Courtesy of SOFTPRO.)

concern about using one's signature as a form of identify verification. In effect, is a person signing for a package delivery actually surrendering a unique biometric identifier to that organization? Or is it simply accepted that the low resolution on the capturing devices used by most courier services are not truly suitable for later verification. How enforceable are signatures taken on these devices anyway?

Another consideration regarding DSV as a biometric identifier is that one's signature means more than just verification, it is used for binding legal authorization and it conveys an authoritative intent. This is the only biometric method that has this dual property of identity verification and authority. Depending on one's point of view, that is not necessarily a bad thing.

Keystroke Dynamics

Keystroke dynamics is a behavior-based biometric technique that verifies the identity of individuals by the manner and rhythm used to type into a keyboard or a keypad (see Figure 2.11). This biometric system develops a unique biometric template using the user's keystroke cadence by measuring and recording such items as typing speed, dwell time, and flight time.

Dwell time is the duration of time in which a key is pressed. Flight time is the time between one key down and the next key down, or the time between a key up and the next key up. The measurements are recorded and processed to create a unique template.

FIGURE 2.11 Keystroke dynamics.

Keystroke dynamics recognize patterns of typing that are based on statistical analyses. The raw data can be captured continuously during a typing session to help determine if the keystrokes match the recorded template. Keystroke recognition biometrics is generally considered to be a very easy biometric technology to implement and use. It is 100% software based. Keystroke recognition requires no additional hardware other than a basic personal computer with which to read, scan, view, record, or interrogate the requesting user because most computers are equipped with keyboards. To authenticate an individual, keystroke recognition relies solely on software, which can reside on a client or on a host system.

To create an enrollment template, the individual might type his or her user name and password a number of times, without making corrections, or he or she might type a preselected text. The best results are obtained if enrollment occurs over an attenuated period rather than at one sitting because individual characteristics are identified with greater accuracy over time. If keystroke errors are made, the system will prompt the user to start over. Some of the distinctive characteristics measured by keystroke recognition systems are:

- Length of time each key is held down
- Length of time between keystrokes
- Typing speed
- Tendencies to switch between a numeric keypad and keyboard numbers
- Keystroke sequences involved in capitalization

Each individual characteristic is measured and stored as a unique template that may be used as a primary pattern for future comparison. Some systems authenticate only upon sign-on, whereas others continue to monitor the user throughout the session.

Disadvantages include variance in typing style during the day or on multiple days, depending on one's state of mind, health, keyboard familiarity, typing duration, and interruptions (e.g., telephone calls or visitors).

Often, the use of simple rules based on a comparison of the live session to the reference session can determine if a legitimate user is attempting to log on to the system. As in other biometrics, the user's keystroke sample is compared with the stored template, and access is granted if the submitted sample matches the template according to preestablished probabilities. This biometric is used primarily in applications where the user is attempting to obtain logical access information from a network or an online system, such as content filtering and digital rights management.

Voice Verification

Voice verification systems (VVSs), also called speaker recognition systems, identify individuals based on vocal characteristics such as pitch, cadence, and tone. Not unlike facial recognition, voice verification is a very natural means by which humans recognize one another. Although I have placed voice verification in the behavioral category, as many biometric professionals do, an individual's speech is influenced by both the physical structure of an individual's vocal tract as well as the behavioral characteristics of the individual.

Voice biometrics works by digitizing a profile of a person's speech to produce a stored model voice print. Each spoken word is reduced to segments composed of several dominant frequencies. Each segment has several tones that can be captured in a digital format. The tones collectively identify the speaker's unique voice print, which is stored in a database in a manner similar to the storing of other biometric data.

To ensure a good-quality voice sample, some voice verification systems require the subject to recite some sort of text or "pass phrase," which may need to be repeated several times before the sample is analyzed and accepted as a VVS template. When a person speaks the assigned pass phrase, certain words are extracted and a sample template is created. When an individual attempts to gain access, his sample voice template is compared with the previously stored voice template. Other voice verification systems may not rely on a fixed set of enrolled pass phrases to verify a person's identity. Instead, these more sophisticated systems recognize distinctions among the voice patterns even when the people speak unfamiliar phrases.

The advantages of voice verification include high public acceptance, no contact requirements, and commonly available sensors such as telephones and basic microphones. It can support remote identification through an existing public telephone system. Moreover, the technology is relatively inexpensive, as it tends to use off-the-shelf, universally available PC microphones or telephones; and it is easy to use and is convenient.

Multiple people have similar voices and all voices vary over time. A person's speech is subject to change depending on health, stress, age, and emotional state. A user must speak in the same voice and tone that was used when the template was created at enrollment. If the person suffers from a physical ailment, such as a cold, or is unusually excited or depressed, the voice sample submitted may not match the template. Background noises and the variation in quality of the input device (the microphone) can create additional challenges for voice verification systems. Even the use of a cell phone instead of an analog phone can impact the accuracy of the results. Moreover, voice verification systems can be somewhat vulnerable to replay attack (e.g., a recording), though some more-sophisticated systems use liveness testing to determine that a recording is not being used. Unfortunately, VVS is plagued with a wider array of environmental and individual variables than any other biometric, including facial recognition.

Voice verification is not sufficiently distinctive for identification over large databases. However, the technology is continually improving its accuracy rates, and it is being successfully used in applications where one's voice must be verified over the telephone such as call-center applications, or to access voice mail, or to deter cellular telephone fraud. The U.S. PORTPASS* program, deployed at remote locations along the U.S.-Canadian border, recognizes voices of enrolled local residents speaking into a handset. This system enables enrollees to cross the border when the port is unstaffed. Voice verification systems have a high user-acceptance rate because they are perceived as less intrusive and are one of the easiest biometric systems for people to use.

NONTRADITIONAL BIOMETRIC SYSTEMS

The human body offers a rich array of uniquely detailed characteristics that can be measured and used for individual authentication, including: body odor recognition, ear-shape recognition, facial thermography, gait recognition, and DNA (deoxyribonucleic acid). These newer biometric technologies use diverse physiological and behavioral characteristics, and they are in various stages of development for future commercial applications.

Body odor recognition uses odor-sensing instruments to capture chemicals that are emitted by skin pores all over the human body. Although it may one day be feasible to distinguish one person from another by odor, the fact that personal habits (such as the use of deodorants and perfumes, as well as diet and medications) strongly influence human body odor renders the development of this technology quite intricate and impractical for most applications.

* *PORTPASS* (PORT Passenger Accelerated Service System) is a generic term for preinspection technology programs developed to reduce the processing time of legal travelers at ports of entry.

Ear-shape recognition is based on the distinctive shape or geometry of each person's ears by measuring the structure of the largely cartilaginous area of the outer ear. Although ear biometrics appears to be promising, no commercial systems appear close to emerging, and ear shape recognition remains a research area.

Facial thermography detects heat patterns emitted from the skin that are created by the branching of blood vessels. These thermogram patterns are highly distinctive. Thermography works much like facial recognition except that an infrared camera is used to capture the images of blood vessel patterns. The advantages of facial thermography include: (1) it is not intrusive since no physical contact is required, (2) every living person presents a usable image, (3) the image can be collected on a moment's notice, (4) a facial thermogram can be captured when there is limited illumination or even in complete darkness, and (5) thermograms can detect certain kinds of disguises. Unfortunately, facial thermography remains relatively expensive, but the technology is moving rapidly toward mainstream commercialization.

Gait recognition recognizes individuals through their distinctive walk by capturing a sequence of images to derive and analyze motion characteristics. Generally, this biometric modality is not significantly impacted by the speed of a person's walk. A person's gait can be hard to disguise because a person's musculature essentially limits the variation of motion, and measuring it requires no contact with the person. It is a promising identifier to recognize people at a significant distance via video images, something other biometric modalities cannot do. On the other hand, the terrain or walking surface can affect one's gait, as might certain attire such as long or bulky coats or special footwear. Preliminary results have confirmed its potential, but not its commercial success, and further development is necessary before its limitations and advantages can be fully assessed.

Currently, DNA matching is not generally considered a biometric because the process is not an automated one; however, this may change in the near future. Today, DNA is mainly used in forensic laboratories, as it does not allow a real-time identification. It is a highly accurate technique where exclusions are absolute and matches are expressed as a probability. DNA enrollment is always possible, but DNA identification is expensive, time consuming (several hours), and needs skilled human intervention. It is also not yet possible to distinguish between identical twins as vein pattern or iris recognition can do.

As has been shown, each biometric technique is accompanied by a set of advantages and limitations. Choosing a biometric technique depends on the application. One must evaluate if the benefits of a given biometric solution offset its limitations and cost.

COMPARISON OF TRADITIONAL BIOMETRIC TYPES

Table 2.1 lists the primary physiological and behavioral biometric modalities in use today with their known advantages and limitations. Whereas Table 2.1 focuses on the characteristics of each biometric modality, Table 2.2 compares the traditional physiological and behavioral biometric modalities against the original seven criteria for what makes a good biometric. In fairness, this is a subjective evaluation.

Table 2.1 Biometric Modalities

Biometric Modality	Advantages	Limitations and Considerations
• Dynamic signature verification	• Readily integrates into e-business applications • An accepted biometric in banking and financial applications	• Accuracy is affected by one's emotional state, fatigue, or illness
• Facial recognition	• Can operate without user interaction • Only current technology capable of identification over distance • Leverages existing image databases • 3D offers increased precision as images capture surface texture of the face on three axes • Can provide a record of potential imposters • Easy for humans to verify the results	• Lighting, variations in pose, and camera inconsistency can reduce matching accuracy • Changes in physiological characteristics or obstructions by hair, hats, glasses, etc., reduce matching accuracy • 2D is susceptible to high false match rates • Faces change over time • Potential privacy concerns
• Fingerprint/ palm print	• Most widely used biometric system • Ability to enroll multiple fingers • Generally uses small, low-cost readers • Uses moderate storage space for templates • Many vendors to choose from • Can be highly accurate • An effective biometric for large-scale systems • Widely accepted forensic tool	• Negative public perception regarding the criminal association (e.g., law enforcement and forensics) • Impaired, dirty, dry, or damaged fingers/palms affect use • Not privacy enhancing; high level of latency • Not considered hygienic • Approximately 2% to 5% of a given population cannot be enrolled • Some fingerprint systems do not achieve the accuracy level that this biometric modality is capable of

(Continued on next page)

Table 2.1 (Continued) Biometric Modalities

Biometric Modality	Advantages	Limitations and Considerations
• Hand geometry	• Operates well in challenging environments; easy to capture • Widely used; established • Biometric is considered highly stable • Very low storage requirements for its templates	• Not accurate for moderate to large populations; human hands are not unique • Readers are bulky and can seem complicated • Perception of passing germs; unhygienic • Not intuitive; requires training
• Iris recognition	• Considered the most accurate modality • High stability of characteristics over time • Hands-free operation • Moderate data storage requirements for templates • Works well with either verification or identification applications	• Moderately expensive implementation costs • Requires more training and attentiveness than other biometrics • Can be obscured by eyelashes, eye lenses, and reflections from the cornea
• Keystroke dynamics	• Very ease to use and to implement • No additional hardware required	• Only useful for applications that require keyboarding and with users capable of using keyboards
• Retina recognition	• Among the most accurate of biometrics • Moderate storage requirements for templates	• Considered intrusive; public has never warmed up to it • Not usable in populations with high incidence of eye disease (e.g., elderly) • Can have high cost; special hardware is required • Limited commercial availability

Table 2.1 (Continued) Biometric Modalities

Biometric Modality	Advantages	Limitations and Considerations
• Vein pattern recognition	• Highly private; no properties of latency • Highly secure; not possible to lift or steal the vein pattern • Very accurate • Small to moderately sized readers • Near contactless, hygienic • High level of usability • Difficult to circumvent • No cultural stigmas to overcome	• Newer biometric; not yet widely used • Can be impacted by bright ambient light and some anomalies such as tattooing
• Voice verification	• High public acceptance • Readily available component parts (e.g., microphone, etc.) • Contactless • Relatively inexpensive; uses commonly available sensors (e.g., telephones, microphones)	• Not sufficiently distinctive for identification over large databases • Generally large storage requirements for templates (e.g., 2 Kbytes to 10 Kbytes) • Use is limited to those applications in which one's voice is being verified • Somewhat difficult to control sensor and channel variances • Influenced by temporary circumstances such as a sore throat, cold, or similar illnesses

Table 2.2 Comparison of Traditional Biometric Modalities

Modality	Acceptability	Resistance to Circumvention	Collectability	Permanence	Performance	Uniqueness	Universality
DSV	H	M	H	L	L	M	L
Facial	H	L	H	L	L	M	H
Fingerprint	M	L	M	M	M	H	M
Hand geometry	M	L	H	M	L	L	H
Iris	M	H	M	H	H	H	H
Keystroke	M	L	M	L	L	L	L
Retina	L	H	L	H	H	H	H
Vein pattern	H	H	H	M	M	H	H
Voice	H	L	M	L	L	L	M

Note: L, Low; M, Moderate; H, High.

3

Anatomy of a Biometric System

Any pattern recognition system that authenticates a user by determining the authenticity of a specific physiological or behavioral characteristic is basically a biometric system. With so many differing biometric modalities, it would seem that each biometric system supporting those modalities would be unique. However, biometric systems have much in common with one another. The biometric components are generically similar in terms of function. Additionally, the biometric system stages—enrollment, verification/identification, and updating—are fairly consistent across biometric types. Moreover, all biometric systems share similar concerns with regard to acceptance, fraud, data storage, and privacy.

Biometric samples are not matched from raw data. Biometric systems acquire raw data from which they extract key features, which are then digitized, compressed, and encrypted to produce templates. A sample template is stored and compared to a reference template that was created during the enrollment process. This is an important privacy aspect of which much of the public remains unaware. The templates that most biometric systems store are simply digitized representatives of one's biometric traits. In most non-law enforcement applications, there are no repositories of individual biometric traits.

This chapter provides a detailed description regarding how the biometric templates are created, processed, and stored. It presents the stages of the biometric process: enrollment, verification or identification transactions, and updates. Additionally, there is a discussion regarding the role of smart cards as the preferred storage mechanism for reference templates.

COMPONENTS OF A BIOMETRIC SYSTEM

The actual number of discrete components in a biometric system varies with each system, and there are a variety of ways to segment a generic biometric system. However, most biometric systems consist of some variant of the following six components or subsystems: (1) sensor/data capture, (2) feature extraction, (3) data storage (also called template storage), (4) matching algorithm, (5) decision process, and (6) administrative subsystem.

Sensor/data capture acquires a sample of an individual's biometric characteristics (for example, an image or signal) needed for recognition, and then converts the sample to a digital format. This process is also referred to as biometric presentation. The quality of the sensor has a significant impact on the system results. Indeed, one could argue that the sensor is the most important component of a biometric system. Biometric data collection takes place during one's registration or enrollment, and precedes verification or identification transactions. The biometric data that is captured should be high quality because poor resolution can produce false negatives and may result in re-enrollments.

Biometric traits can be collected in various ways. A photograph can collect data for facial recognition, and a recorder can capture a quality voice print. The captured data is referred to as raw data. Consistency in the manner in which a sample is collected is critically important. Changes in sample collection can affect the accuracy of the template, which in turn affects the probability of a match. The sensor output is sent to the signal processor to *select* and to extract the distinguishing characteristics of the sample.

Feature extraction automatically processes the acquired information by extracting only the key data from the distinguishing features of a biometric sample, generating a digital representation referred to as the biometric template, and performing some quality control checks. This may involve a feature extraction, or it may require locating the biometric characteristics within the received sample. This is where the signal processing algorithm comes into play. Predetermined, discriminative specific features are extracted from raw data to form a new representation of only system-significant information; the rest of the raw data is discarded. This new representation should be unique for each individual as well as be somewhat invariant with regard to multiple samples garnered from the same person over time. The biometric system then digitizes, compresses, and encrypts those features to produce a template. It is this template, not the raw data, that will be used in the matching process.

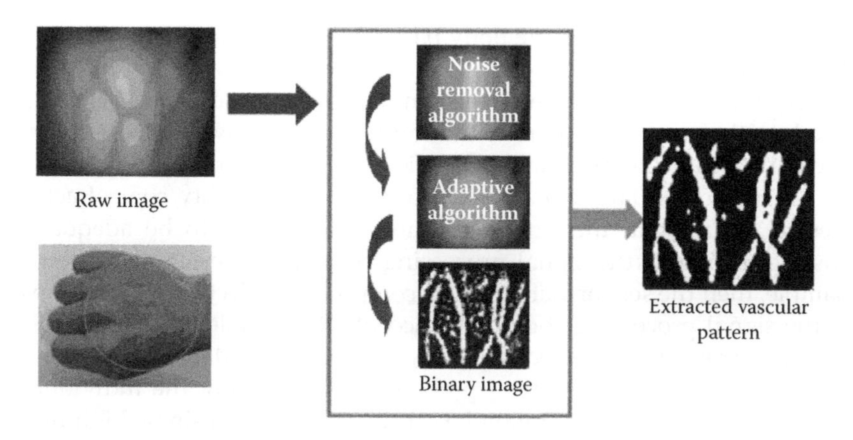

FIGURE 3.1 Vein extracting processing flow. (Courtesy of Identica Corp.)

Feature extraction might be viewed as nonreversible compression. The feature extraction is executed during the enrollment process as well as during the authentication process. During enrollment, the extracted biometric data is used to create a reference template with associated personal data. In the authorization process, the feature extraction is applied to each frame of the scanned image prior to the matching process with the enrollment template.

As Figure 3.1 illustrates, the back of the hand is imaged by the vein pattern recognition (VPR) camera sensor, and a raw image is extracted. The extraction process selects the distinguishing characteristics from the raw data sample, converts it to digital format, and then processes the digitized data into a compact biometric identifier record—the biometric template. Generally, biometric templates require much less memory for storage than the biometric data samples. Raw data is simplified through the process of feature extraction. After the noise removal and adaptive algorithms such as contrast enhancement are applied, the biometric system creates a binary image of the biometric representation, and then it stores that extracted data. The readily visible patterns of veins on the back of the hand are an example of distinguishing characteristics for vein pattern recognition biometrics. At the completion of the extraction process, it is virtually impossible to reconstruct the original hand scan from the digitized template.

The selection of the biometric features is highly dependent on the extraction algorithm. Consequently, the features extracted by one algorithm can differ markedly from the features extracted by another algorithm, even for the same body part. The implication of this reality is that a template produced by one biometric system may not be fully compatible with another biometric system of the same modality even when used on the same body part of the same person. However, the template incompatibility situation can be beneficial from a security standpoint.

Extraction usually entails performing several quality control activities to ensure that the extracted features are likely to be adequately distinguishing. If the signal processing component rejects the received sample, then the sensor/data capture component collects another sample. If the signal processing component accepts the sample, it then creates a template from the extracted data. Depending on the deployed biometric system, during the enrollment process multiple sets of the individual's biometrics characteristics might be captured to create a single high-quality enrollment template, or multiple templates are created and stored to account for *intraclass variations*.

Intraclass variations are situations in which a user's live data sample will be somewhat dissimilar to the user's template created at enrollment. The cause of the variations might stem from differing interactions of the user with the sensor (e.g., inconsistent placement or pressure), changes in the environmental conditions (e.g., lighting variation), use of different sensors during enrollment and at verification, changes in pose, or changes in the biometric trait (e.g., illness or aging). Such situations lead to a high false rejection rate. On the other end of the spectrum is interclass similarity. With interclass variation, the randomness of a data pattern among individuals is generally the norm, as measurable traits can vary significantly; and that contributes to low false acceptance rates. However, in some isolated cases, rather than variation, these large interclass similarities in the feature sets limit the discriminatory ability expected for that trait. High levels of interclass similarity lead to higher false authentication rate.

Template storage houses the enrollment templates to which the new biometric templates will be compared. Templates are stored within an enrollment database. Templates may be stored within a biometric capture device, on a portable medium such as a smart card, in a distributed device such as a personal computer or local server, or in a central repository. The templates can be stored in an altered format, compressed, and encrypted.

The choice of template storage may be determined by the biometric method used, by the level of privacy required, by the choice of biometric devices, and by the enterprise business model. The amount of data storage needed for a given template varies by the biometric modality used, the specific system deployed, and the application supported. Since templates require only modest memory, ranging from less than 15 bytes (e.g., hand geometry) to around 3,000 bytes (e.g., facial or speaker recognition) in size, storage space is not typically a major issue except in large implementations. There are typically four options for reference template storage:

1. Locally store the template within the biometric reader itself or in another localized database. This enables a fast response during future verification transactions. The key advantage is that this method eliminates various points of failure, as there is no need to communicate to a central database. However, in scenarios where individuals need to access multiple locations across a given geography, the database must be replicated at each reader. This may require extensive memory within the reader, and it imposes the requirement to update each reader with a predetermined frequency.
2. Locally store the template at individual workstations. The workstation creates less privacy concerns and provides a basic level of security, although placing the biometric template on the hard drive may not be the most secure. Additionally, workstation storage limits verification applications to the workstation itself, as this solution is not generally appropriate for multiple location use.
3. Remotely store the template in a central data repository. For identification applications the sample templates must be compared to the entire template repository; in these identification scenarios, a central database is often the best storage solution. However, for distributed verification applications, access to a central template repository would heavily depend on the data network that connects it to the reader device; and a network would introduce multiple points of failure. It would be vitally important that all transmitted data be encrypted to counter fraudsters who might otherwise "sniff" the biometric data off the network and replay the authentication session in an attack. In addition, some users are very privacy conscious and do not like the idea of their enrollment templates being stored centrally.
4. Securely store the template on a portable token such as a smart card. This method addresses many of the drawbacks of previous methods, except for identification applications. The biometric data is not centrally stored, does not traverse the network, and individuals carry their biometric templates with them. Many individuals

prefer this method since it gives them a sense that they control their personal identification data. The one drawback is that the cost of the biometric implementation is marginally higher, since smart cards and smart card readers would also be required. However, both cards and card readers are relatively inexpensive. Indeed, new smart card form factors have been developed such as universal serial bus (USB)-based smart tokens for which no external reader is required. In most applications smart cards are the container of choice for biometric templates because they offer greater mobility, flexibility, and security; have fewer points of failure; and allow for better support of privacy goals. Furthermore, many biometric vendors are now building smart card support into their biometric readers.

Matching algorithm compares each new sample template to the stored reference template and matches the similarities. The analysis is then passed as similarity or match scores to a decision processor. The similarity scores indicate the degree of fit among the templates compared.* With verification applications, a single specific claim of subject enrollment results in a single similarity score, which is usually a quantitative probability estimate. With identification applications, all reference templates in the database may be compared with the subject's sample template, and the outputs include a similarity score for each comparison. However, biometric systems only present a likelihood of a match as expressed in a probability; and despite the fact that these likelihoods are relatively accurate, they are not absolute.

For larger biometric populations, response time becomes very important, and it is highly dependent on search and retrieval accuracy. Biometric data does not have a natural sorting order; and the computational overhead of pattern matching for large-scale identification systems can be extensive, requiring hours or even days of processing. Therefore, biometric practitioners have had to devise techniques to index biometric information to reduce that overhead and, consequently, to accelerating the matching process. One well-accepted method is to accomplish this is by classifying the biometric data. Thus, binning may come into play.

* Similarity scores represent the similarity of the sample template under consideration with the reference template. Sometimes it is referred to as a genuine score if two samples of a biometric trait match, but it would be called an imposter score if the two samples do not match.

Binning is the process of classifying biometric data to enable the presorting of very large biometric databases. This technique can significantly speed the effort in matching sample biometric data with the reference template. In binning, reference templates are partitioned into characteristics or bins. The bins selected for classifying data can be based on external characteristics (such as gender) or on internal characteristics (such as whorls, loops, and arches in the fingerprint modality). Indeed, many traditional fingerprint identification applications are based on binning. Binning is very useful in accelerating identification processing, and it enables more accurate statistical matches. Nevertheless, all such classification techniques can introduce hidden obstacles—if the reference template has been inaccurately categorized, it becomes very difficult to obtain a correct match.

A *decision processor* uses the scores of the matching component to make a system-level decision for a verification or identification transaction. Unlike a password-based system in which the submitted code is either correct or it is not, biometric matches are based on probabilities within intervals of confidence. A newly created sample template is generally not an exact match to the reference enrollment template due to an array of variables. Of course, there is usually some tolerance involved, but biometric systems can differ significantly regarding the amount of variance that can be accepted for matching. Most biometric systems establish a threshold that must be exceeded for a match to occur. Therefore, a threshold should be set at a level to accept all similarity scores that are considered matches but high enough to reject fakes. A decision is rendered based on the match score and a predetermined threshold of parameters for acceptance or rejection. Most biometric systems base their decisions on statistical methods, whereby the average of several samples is calculated, and a normal distribution of the biometric data with a mean value and standard deviation is formulated. The quantitative difference between the sample and template must lie within a certain threshold to be accepted as a match, or it is rejected.

Templates are considered a match when the similarity score exceeds a preselected threshold. The individual's claim can then be verified on the basis of the decision policy. This decision process can be fully automated or it can support human intervention. The decision policy may allow or require multiple attempts before making an identification decision.

An *administrative subsystem* governs the overall procedures and usage of the biometric system. Not all biometric practitioners would list an administrative subsystem as one of the key system components. However, there are a number of administrative functions that do take place, regardless of how someone categorizes them. Examples of administrative functions include:

- Providing feedback to the subject during or after data capture.
- Storing and formatting the biometric templates and biometric interchange data.
- Providing final adjudication on output from decisions and scores.
- Maintaining threshold values.
- Maintaining biometric system acquisition settings.
- Managing the operational environment.
- Interacting with the application that uses the biometric system.
- Providing preselected or ad hoc reporting.

STAGES OF THE BIOMETRIC PROCESS

The previous section describes the biometric process from a system perspective. The following section explains the same processes from a user perspective. It is instructive to segment the generic biometric processing experience into three key stages: (1) enrollment, (2) ongoing transactions (both verification and identification), and (3) updates (re-enrollments).

Enrollment registers an individual to a biometric database for the first time. This is the process in which a user presents a sample for the biometric system to convert to a reference template and to store it in the system database along with a proxy for his or her identification. The user's initial biometric sample is collected, assessed, processed, and stored. To a large degree, enrollment helps determine the ultimate accuracy of future matches. To ensure accuracy, a good enrollment technique is to take three separate biometric measurements. Enrollment can take place for both verification and identification applications. However, most biometric systems cannot support both verification and identification simultaneously, although both applications can share the same biometric database.

In the enrollment process, an individual submits behavioral or physiological data in the form of biometric samples to a biometric system. A usable submission may require looking in the direction of a camera or placing a finger on a biometric reader. Depending on the biometric system, a user may have to remove eyeglasses, hold his hand still for a second or two, or recite a predetermined phrase in order to provide a biometric sample. Typically, an enrollment includes the following steps:

- Submission of a biometric sample.
- Segmentation and feature extraction.
- Quality assurance checks in which the system may reject the sample or features as being unsuitable for creating a template, and require acquisition of further samples.
- Reference template creation (which may require multiple samples).
- Potential conversion of a template in a data interchange format and storage.
- User test of a verification or identification attempt to ensure that the resulting enrollment is usable.

Generally, no two reads from any biometric reader are exactly the same, and thus no two templates from the same person are exactly the same. Thus, there is a small element of uncertainty. Small variations in one's positioning, the distance from the sensor to individual's identifying feature, the subject's interface with the sensor, and environmental conditions (humidity, temperature, sunlight, and so forth) each contribute to rendering a template unique. Each time biometric data is extracted, it is used to create another unique sample template for comparison to the reference template. The quality of each template—both the reference template and each subsequent sample template generation—is critical to the success of the overall system. The better the quality of the templates, the more accurate will be the overall system and the lower the error rate. Discarding imperfect images created during enrollment should actually improve the overall accuracy of the system. Figure 3.2 graphically illustrates the subprocesses involved in enrollment.

During enrollment most biometric systems collect ancillary information about the enrollees. This ancillary information may include

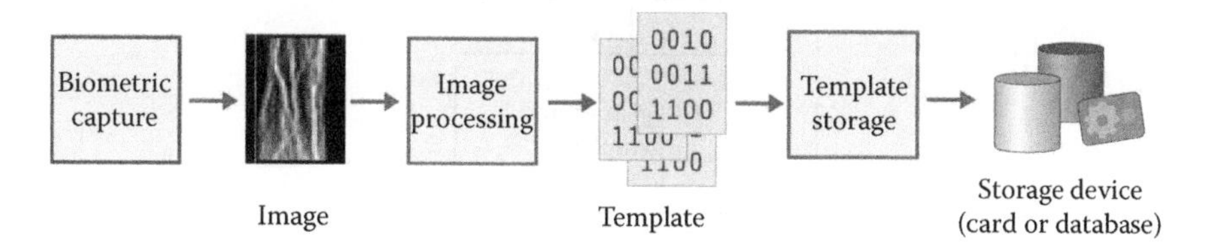

FIGURE 3.2 Enrollment process.

demographic information (e.g., organization, division, address) or soft biometric information (e.g., gender, age, eye color) to help in verifying identity. It is very important for the organization collecting this information to securely store and safeguard it, partition it for specific uses, and to delete it when the individual is no longer participating in that particular endeavor.

Enrollment is the most crucial stage of the biometric process. Nothing can influence the successful use of the biometric system more than enrollment. If one fails to properly capture the biometric trait and create an accurate reference template, then subsequent attempts to match templates will fail and that will continue through the biometric life cycle for that individual. The best opportunity for success is to establish a strong enrollment process and for individuals to follow enrollment procedures precisely. A weak process may lead to system inaccuracies and inconsistencies, as well as inadvertently create an unreliable authentication infrastructure. Additionally, an improper enrollment, whether intentional or inadvertent, tends to militate against the purpose of biometric security, and it creates a compromised system that can be very difficult to correct. For many users the enrollment phase may be their initial exposure to a biometric system. Thus, it is critically important to provide users with specific and proper instructions, and to enable users to offer their feedback and to ask questions.

During the enrollment process a user record is created. That user record consists of an identifier and the biometric template, both of which should be encrypted (see Figure 3.3). The identifier links the template and

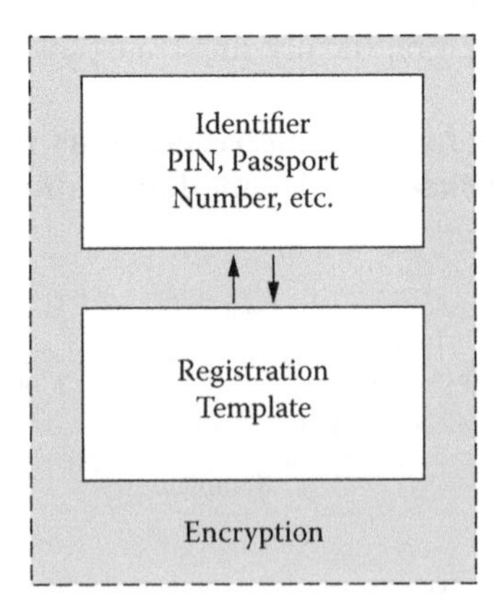

FIGURE 3.3 User record.

the user. A single user can be associated with various identifiers, which may lead to multiple roles in the biometric system for the attribution of rights and privileges. Thus, the biometric template and user credentials are bound together via identifiers. This also prevents an attacker from taking someone else's user credential by attempting to link his identifier to it. To do so, the attacker would need a valid user credential already on the biometric system, and if he had such a credential, such an attack would most likely be unnecessary.

The biggest failure of biometric systems is ensuring individual identities at the time of enrollment, as there is no absolute guarantee to that identity. Each enrollee usually presents some background information that vouches for his or her identity, but fraudulent information such as a fake or stolen birth certificate or driver's license might be presented by a fraudster. For this reason it is crucial for the enrollment process to rely on multiple valid methods of identity proof.

To ensure integrity of a secure enrollment process, the enterprise or other enrollment authority should require comprehensive identity authentication (e.g., multiple IDs) prior to an individual beginning the enrollment. There are several other techniques that can be used to help ensure the integrity of the enrollment process. One test is to perform a matching process of each new enrollment against the existing database to check for duplicate entries. For many applications, an identification process is used only at the time of enrollment to verify that the individual is not already enrolled. The biometric identifier only verifies that the person is who he says he is at enrollment.

The biometric system should always store in its database a record of the enrollments, and successful and failed verification attempts. Storing the confidence levels achieved at each verification is also advantageous, since they can be used later to determine when a feature is moving out of tolerance levels.

Most technologies that are employed to solve a set of problems often bring with them some issues of their own. Biometric technology is no different. There are two key issues regarding enrollment that must be considered before the system is fully deployed: failure to enroll situations and potential circumvention.

Potential enrollment problems exist with each biometric modality, and there are trade-offs that must be addressed. In some cases, it means simply matching the appropriate biometric type to the application. Given a large population of users, regardless of the technology, some small percentage will be unable to enroll since no biometric modality can claim to work for 100% of a sizeable population. Simply put, all biometric systems require a backup, which is often a manual process.

Even though vein pattern recognition can be used by over 99.8% of most populations, there are still those who fail to enroll due to a physical issue or for some other reason. What happens if a person cannot be enrolled? This is due to many factors including an individual's lack of the requested biometric identifier. And sometimes people just have an off day. It is important for the enterprise to plan for and implement alternatives for these individuals. This is referred to as exception handling, and all biometric systems will have an occasional need for it. The situation might be handled by the implementation of another biometric modality, or it could be resolved through manual procedures, or invoke some other remedy, depending on the application. When that situation occurs, the organization sponsoring the biometric system simply needs a predetermined immediate action drill that it can activate to resolve the exception. From both a systems integrity and fairness standpoint, the exception handling should always be consistent, thorough, and each event well documented.

For reasons that were alluded to earlier in this chapter, the best way to circumvent a biometric system lies in using false identification at enrollment. That is why a biometric system is only as secure as its enrollment process, which depends, in turn, on the quality of the personal identifiers that the enrollee presents. Most personal identifiers that would authenticate one's identity prior to enrollment tend to be document based. The documents presented to verify the enrollee's identity need to provide iron-clad authentication of the enrollee's true identity. A secure biometric enrollment process is one that establishes with nonrepudiation an individual's identity and which determines what privileges are being granted to that person. To deter impersonation and to ensure the privacy of each person's credentials, an enrollment process should implement appropriate physical security measures at the enrollment site. Moreover, every biometric system should have supervised in-person enrollment. Spoofing biometrics is discussed in greater detail in Chapter 6.

Verification/Identification/Updating Transactions

Security systems use biometrics for two basic purposes: for identity verification (also called authentication) and identification of individuals. Once enrollment and storage are complete, users authenticate themselves by matching live sample data to the reference template. Alternatively, a user sample might be fed into the biometric system to determine the identity of the sample (e.g., identification). Updating adjusts the reference templates. The following section describes the processes for verification, identification, and updating.

Verification is any application that requires individuals to authenticate a claimed identity. Moreover, verification activities tend to be self-service based, transaction-oriented, and unsupervised. The biometric system captures an individual's biometric image, and then extracts the unique characteristics from the individual's image to create the user's sample template, also referred to as a trial template or a live template. The biometric system then compares the sample template to the template stored at enrollment (for example, the reference template), and in most systems, a numeric matching score is generated based on the percentage of similarity between the sample and reference templates. Depending on the predetermined threshold value, the identity verification score may or may not meet the probability threshold for a match. This process is also referred to as one-to-one (1:1) matching.

Figure 3.4 illustrates a standard biometric verification process. For larger systems, the user enters a personal identification number (PIN) and then provides his or her biometric sample (e.g., a finger vein pattern). PINs serve as record locators and may not be required in smaller implementations. Comparison of sample data and reference templates results in a biometric system match decision; that is, a yes or no match. Verification biometric systems bind the sample to the template for a one-to-one match, or no match.

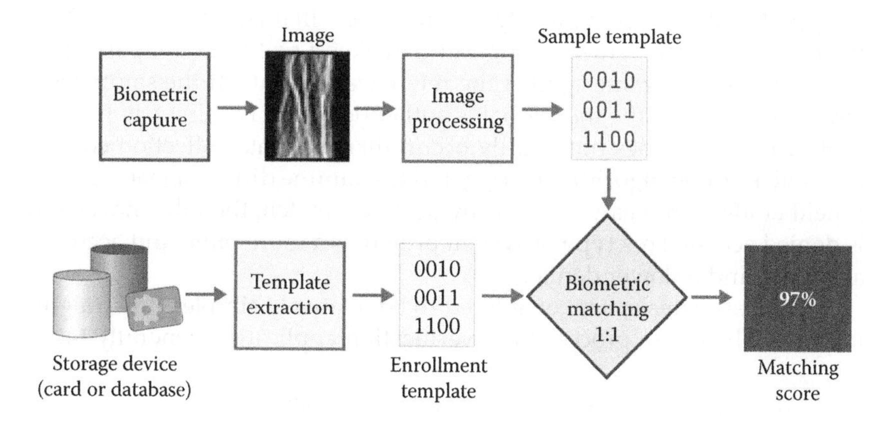

FIGURE 3.4 Verification process.

The verification decision outcome is considered to be erroneous if either a false claim is accepted (false accept) or an authentic claim is rejected (false reject). Note that many biometric systems will allow a single end user to enroll more than one instance of a biometric characteristic.* For example, a hand vein system may allow end users to enroll both hand images (palm or back of the hand), while a finger vein pattern recognition system enables end users to enroll multiple fingers, in case some fingers or a hand is injured.

Typically, a verification transaction includes the following steps:

- Submission of a biometric sample.
- Segmentation and feature extraction.
- Quality assurance checks in which the system may reject the sample features as being unsuitable for comparison and require acquisition of further samples.
- Comparison of the sample template against the reference template for the claimed identity producing a matching score.
- A review on whether the sample template matches the reference template as it relates to the threshold score (no match is ever perfect because of the relative uniqueness of each template).
- A verification decision based on the 1:1 match result of one or more attempts, depending on system policy.

Some VPR systems such as the ones produced by Hitachi-Omron Terminal Solutions use a multipass comparison algorithm. During the authentication process, the finger vein module continuously captures still images at a rate of approximately 30 frames (or still images) a second. Once the finger is detected, the first algorithm makes a real-time comparison of each new sample frame against the reference template. If the sample is a strong match, then the individual is authenticated in the first few frames. If authentication does not readily occur, then the data collection continues, and a second algorithm is triggered to examine different feature data. If neither algorithm can successfully declare a match, then the individual is denied access. This type of system promotes a more rapid and accurate approval, and a slower denial.

Verification tasks are considered to be relatively simple and straightforward. The entire process for a verification application generally takes

* Instances are records or images of the same biometric identifier.

only a few seconds, depending on the biometric modality employed and the specific implementation. Another advantage is that verification enables unattended access. Some systems allow multiple sample retakes to produce a match. If an individual is denied, human interaction may be needed, and the denied person may need to follow a manual process for authentication.

As mentioned in the previous section, large biometric systems may use PINs as record locators. Throughout this book I have suggested that biometric identifiers are superior to PINs and passwords; now I am suggesting that one use a PIN with a biometric identifier. This may appear somewhat inconsistent to many readers. However, complex passwords that must often be changed are difficult to remember, as opposed to a simple, self-selected four-digit PIN. When the PIN is used with a biometric identifier, it does provide another security factor, but its true purpose is not so much security as it serves as a record locator to efficiently retrieve one's reference template. Rather than search through a large repository of reference templates against which to make the comparison with the live template, the PIN enables the system to quickly locate only those templates with the same PIN. This significantly speeds up the verification process.

It is logical that there would be an overlap of templates with the same PIN since users select their own PINs. That subset of individuals with the same PIN would still be much smaller and more manageable than the larger database. The biometric trait, not the PIN, provides the identity verification. Further, a smart card can securely hold and safeguard many complicated passwords and multiple biometric identifiers, and can increase security and user convenience. Once a user's identity has been confirmed within an acceptable confidence level, he could use the more complex passwords stored on the smart card to access other systems.

Identification applications are those in which no claim to identity is made, but for which there is a need to reveal the identity associated with a biometric trait. The biometric image is submitted to a biometric system, which extracts the unique characteristics it will use to create a trial template and compare it with all known templates in the database. Potential matches may surface with a given probability of a match. Identification validates that an individual exists in the known population, or it confirms that a given individual is not enrolled with another identity or is not on a preselected watch list. This is also referred to as one-to-many (1:n)

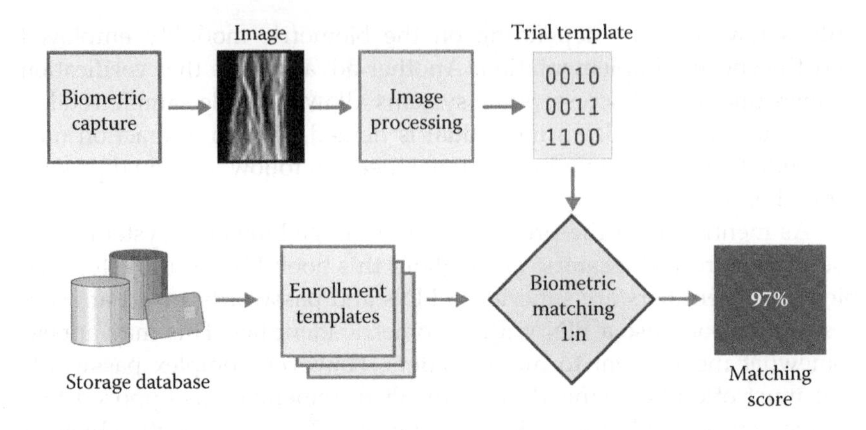

FIGURE 3.5 Identification process.

matching. Figure 3.5 illustrates the identification process for a standard biometric system. Typically, identification involves the following:

- Submission of a biometric sample.
- Segmentation and feature extraction.
- Quality assurance checks in which the system may reject the sample or features as being unsuitable for comparison.
- Comparison against some or all templates in the enrollment database, producing a matching score for each comparison.
- A review on whether each matched template is a potential candidate identifier for the user, based on whether the similarity score exceeds a threshold or is among the highest similarity scores returned.
- An identification decision based on the candidate lists from one or more attempts, depending on system policy.

Identification can occur in two different modes: positive and negative. Positive identification refers to making a determination that a given sample is in the identification database. Negative identification can be used to confirm that the individual is not enrolled with another identity* and is not in a predetermined biometric database. The individual does not claim any alternative identity, but instead he effectively claims that he has

* Enrollment in an identification application may or may not be voluntary. For example, when the Federal Bureau of Investigation (FBI) "enrolls" a criminal or when U.S. Homeland Security includes potential terrorists in their watch list, the individuals in question probably did not consciously and voluntarily enroll in the database.

not already enrolled in the program in question. This is a very powerful mechanism as it prevents an individual from using multiple identities within a given system. It is particularly relevant for large-scale public programs such as government welfare or other benefits programs where, as a precondition for acceptance of the individual into the program, the government must confirm that he or she has not previously enrolled. Its implementation would typically need a secured database containing all applying individuals and their biometric identifiers. The biometric system would create an individual's template and compare that sample template against all the previously stored templates.

Negative identification applications can be used to prevent duplicate records of the same individual. Sometimes individuals may not realize that they were already enrolled; or it could apply to situations where multiple records were spawned from administrative error. As an example, hospitals are notorious for creating multiple patient records because of a typographic error in the name, or the inclusion or exclusion of a middle initial from one visit to the next. Negative identification could help eliminate such errors. State governments in the United States continue to search for ways to detect multiple driver's licenses issued concurrently to the same individual.

In identification, the biometric system seeks an identifier from the subject's enrollment. Identification provides a candidate database of biometric identifiers. Identification is considered correct when the subject is enrolled and an identifier from the enrollment is in the candidate database. The identification is considered to be erroneous if either an enrolled subject's identifier is not in the resulting candidate database (false-negative identification error) or if a transaction by a nonenrolled subject produces a potential match within the candidate database (false-positive identification error).

Whereas individuals use multiple means to authenticate themselves in a verification application, such as through the use of PINs or passwords, smart card tokens, or biometric identifiers, identification applications tend to depend solely on one's biometric identifiers. Verification and identification processes have similarities, but their differences are stark. Figure 3.6 illustrates some similarities and differences between verification and identification processes.

Updating remeasures and adjusts the reference template. Most biometric systems maintain stability (permanence) over time; however, changes can occur. The primary reasons for change include modification in sensor characteristics (e.g., new biometric reader models), variations

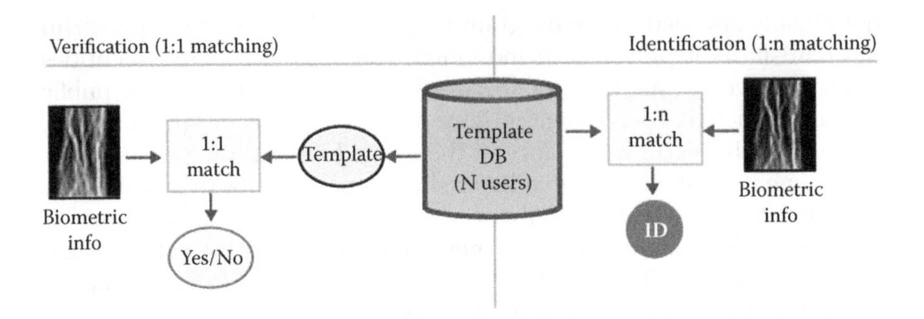

FIGURE 3.6 Verification and identification compared.

in environmental factors, and alterations in the biometric trait itself (e.g., scars, cuts, or disease). It is possible for a reference template to become significantly misaligned. Moreover, if users are experiencing problems with a biometric system, they may need to re-enroll to gather higher-quality data. The user may not have been properly enrolled or some other improper interaction with the enrolling sensor may have occurred.

Periodic updates or re-enrollments help address the instability of some biometric techniques over time. For example, fingerprints are generally highly stable over time, but can be affected by aging or occupational hazards that damage the skin. Iris patterns are stable until death unless obscured by eye disease. Face, hand geometry, and voice recognition are more susceptible to weight fluctuation, aging, and other changes; these biometric modalities benefit significantly from periodic re-enrollments.

There are basically two methods to effect a template update. The first method is a replacement update because the original reference template is discarded and is replaced with a more current template. It is similar to a re-enrollment process. The second method builds on the original template and constantly adds additional data sets with each new transaction. This is an augmentation update.

As with the initial enrollment process, it is very important to maintain the integrity of a secure process by making enrollees prove their identities, preferably through multiple means. The information used to re-enroll or update an individual must be of the highest quality. As with the initial enrollment process, it is a good practice to run an identification check against one's template database to ensure that the enrollee has not already enrolled as someone else.

FIGURE 3.7 Smart card reader attached to a PC.

BIOMETRICS AND SMART CARDS

A smart card is often a key component of an access system. Enterprises benefit from the enhanced layer of identity verification (for example, a physical token); smart cards support strong machine-generated passwords that they will not "forget"; and they provide secure, portable storage for digital certificates and private keys. Smart cards also simplify the user experience by enabling the user to support multiple security systems across an enterprise through the use of a mobile smart token, access to which occurs simply by inputting a single PIN. As shown in Figure 3.7, a smart card reader has a small footprint and can attach to a PC via a USB port. Or the "card" could be a USB smart token that would be inserted directly into the PC. The power and portability of the smart card with its internal card security and strict operating procedures help maintain a reliable chain of trust.*

Once a system digitizes biometric data and stores it on a computer system or on the biometric device itself, then that data would be exposed to the same vulnerabilities typically faced by most computer systems. Alternatively, a smart card possesses unique security features that can protect that data, such as anticounterfeiting and high-level tamper-resistant features including the capability to detect and to react to tampering

* A chain of trust refers to guarantees of authenticity of the people, enterprises, network, equipment, and other components of a secure ID system.

attempts. One key method of protection is its offline capability such that no hacker can reach the data on a smart card when it is in one's pocket, disconnected from any computer system or reader device. Even contactless smart cards require the proximity of a reader, and many require PINs to activate. Additionally, smart cards can store peripheral information that helps manage the biometric program. For example, the card could store successful and failed verification attempts as well as the confidence levels achieved at each verification. Combining smart card and biometric technology enhances security by providing a higher level of assurance, increases the system's flexibility, and reduces the processing time needed to match a given biometric with its stored template.

Template storage on a distributed biometric device, such as a biometric reader or server, could have storage size implications, and it could limit users to only those access points that contain their templates in the local repository. Storing the biometric template on a smart card requires no data repository of templates, either centrally or locally, as the user carries his template with him. It also eases users' fears of having biometric data stolen from or inappropriately accessed from a central database. Of course, loss of the smart card would require re-enrollment, but that would affect only the individual who lost his card. This option better protects the privacy of the user by enabling him to maintain and control his own reference template. The use of smart cards for biometric template storage does incur additional costs, not only for the cards (about \$2 to \$3 per card) themselves, but for smart card readers (about \$20 to \$40 per reader) at each access point. Nevertheless, many early adopters of biometrics have opted to use smart cards because it obviated the need to maintain databases and transferred possession to the individuals.

If a smart card is used to house one's reference template, there are several variations regarding how the smart card and biometric system might interact. The template may be stored on the smart card but then can be transmitted to the host or biometric device for verification applications. This is the standard way that a smart card works with a biometric system. Two corollaries of that variation include (1) the extraction of the key data features for comparison could occur in the biometric device, but the actual comparison and decision making could occur on the card; or (2) the card could be sent the match score, and it would render a decision based on its on-card logic.

Alternately, the sample biometric template could be passed to the smart card to perform a secure match with the stored reference template. On-card matching would ensure another level of security as well as privacy, as the reference template would never leave the smart card, and a comparison of the biometric templates would be performed inside the microprocessor chip. By placing the reference template on a smart card and performing a match-on-card, the individual would always be in control of his or her biometric identifier. For additional security, the templates could include a cryptographic signature signed by the application that generated them.

Another key advantage to the use of smart cards for on-card matching applications is the scalability of the system. There is no limit to the number of individuals who can use on-card matching for verification without any requirements for server processing or other IT resources. On-card matching offers a distributed, mobile, and highly scalable solution that is particularly appropriate for large-scale implementations, especially when thorough database maintenance is impractical.

One concern that has been occasionally voiced is overwriting the card with a fraudulent biometric. However, that is unlikely with today's advanced smart cards. The biometric template would be digitally signed and locked onto the smart card. Smart cards reject modifications to their memories without authentication. If a fraudster attempted to steal the digitized template from a valid smart card, he would not be successful since smart cards are tamper-resistant and have internal logic that detects if they are being hacked. In such cases, they will shut down or even self-destruct, depending on the smart card operating system and the application.

A combination of smart cards and biometrics make a great deal of sense for applications where strong authentication is needed and where users have trouble remembering multiple or complicated passwords.* Smart cards are a vital link in the chain of trust for identity systems; biometrics represents another great tool for verifying identity. Not only does a combination of smart cards and biometrics enhance one's privacy, but it also heightens one's confidence in the verification of an individual's identity. Additionally, the biometric template stored on the smart card can itself be encrypted to improve security against external attacks.

Finally, smart cards are an excellent storage and retrieval token for multimodal biometric systems. In addition to the previously mentioned security features, some smart cards can separate biometric templates and their associated PINs by internal firewalls so that they can be used

* Although a strong password is one that is composed of letters, numbers, and symbols and cannot be guessed (e.g., ksy&9m%n43d), it is difficult for most individuals to remember even a moderate number of simple passwords.

independently. Smart cards and biometrics are a natural combination. A smart card is the logical storage device for safeguarding biometric templates, for performing on-card matching, and for supporting the portability of one's credentials and access privileges. Biometric systems extend the functionality and value of smart cards. Smart cards strengthen security, safeguard biometric templates, and keep them private. Together, biometrics and smart cards are critical components in a potent portable security system.

4

Vein Pattern Recognition Modality

The human vascular structure is individually distinct and appears to be time invariant. Human blood vessels are formed during the embryo stage* with a variety of differentiating features, rendering each pattern unique, and their patterns remain relatively constant over one's lifetime. A unique network of veins and arteries exists in every hand and finger of each human being.† However, blood vessels are not exposed, and the intricate network pattern is usually not observable within the visible light wavelength.

An individual's identity can be authenticated using vein patterns in one's hands and fingers, and those patterns are located just under the surface of the skin and can be clearly mapped with low-cost, high-resolution CCD‡ cameras (see Figure 4.1). The camera sensor of a vein pattern recognition (VPR) device is able to detect and recognize the vein pattern through the hemoglobin that actively flows in the individual's veins;§ those veins

* The embryo stage generally spans the time between the week after conception until the end of the eighth week, when the fetus stage begins.

† As mentioned in Chapter 2, technically retina scanning is a form of vascular pattern recognition; and facial thermal imaging does map vascular patterns in one's face using infrared (IR) light. However, the hands (dorsal side, palm and fingers) are considered the most convenient parts of the human body for vascular pattern measurements; and thus, the terminology name is usually associated with fingers and hands.

‡ CCD is an acronym for charge-coupled device.

§ Hemoglobin is an iron-containing protein pigment occurring in the red blood cells and functioning primarily in the transport of oxygen from the lungs to body tissues. Hemoglobin acquires oxygen in the lungs and carries the oxygen through the arteries to the body tissues. The deoxygenated hemoglobin returns through the veins to the heart and lungs.

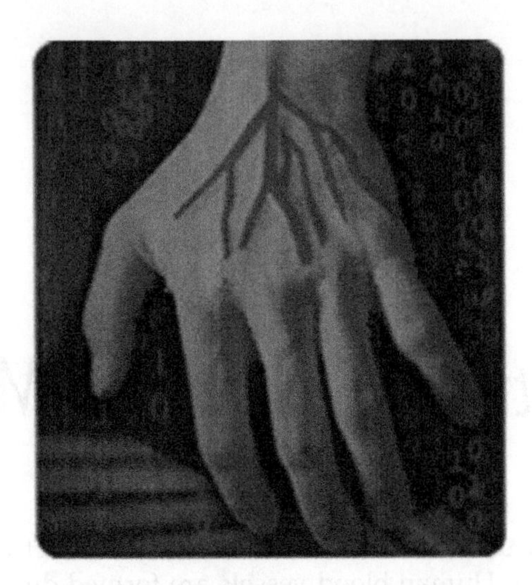

FIGURE 4.1 Vein pattern image. (Courtesy of Luminetx Corp.)

appear as a pattern of dark lines against a light, almost white, background. Vein pattern imaging "visualizes" the vein network patterns by exploiting the optical characteristics of hemoglobin. The absorption images of the hemoglobin in deoxygenated veins differ markedly from the absorption images of oxygenated arteries. The venous blood carries waste such as carbon dioxide and it has a bright red color, and veins in one's hand are located closer to the surface than the arteries. VPR technology further refines the raw pattern images by applying illumination control and by employing advanced contrast-enhancement techniques. As a result, individuals can be rapidly verified against a stored reference template, providing very fast and robust biometric authentication.

VPR technology was developed in the 1990s, and the first commercialized system emerged in 2004. Different vendors focus on the vein patterns located in various parts of the hand—finger, palm, or back of the hand—but each technique shares similarities in approach. Most VPR systems use a set of light-emitting diodes (LEDs) to generate near-infrared light. However, the intensity of the near-infrared illumination is an order of magnitude less than the near-infrared rays contained in sunlight.

Vein biometric systems record subcutaneous infrared absorption patterns to produce unique and private identity verification templates. As depicted in Figure 4.2, the optical unit controls the lighting dependent

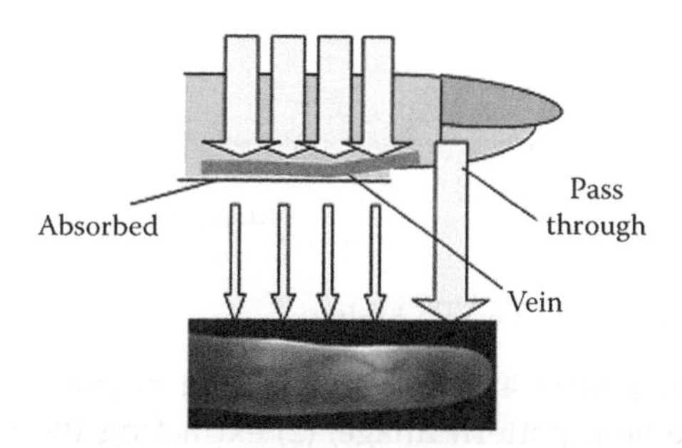

FIGURE 4.2 Near-infrared light imaged by a CCD camera. (Courtesy of Hitachi Ltd.)

on the illumination surrounding it. The light emitted from the LEDs penetrates the finger and enables the CCD camera to generate vein pattern images. The wavelength of the near-infrared light must be adjusted to maximize the contrast between the veins and the finger background. The difference in the absorption of near-infrared light between veins and other body tissue is significant enough to produce a readily discernible sensor image of the vein pattern. After veins absorb the light rays, the veins appear as dark images to the camera, while the smaller, oxygenated arteries are less visible to the camera's sensor. The vein pattern is digitized into a binary form, and then the data is extracted, encrypted, and stored as a template.

VPR AUTHENTICATION PROCESS

VPR systems generally consist of a near-infrared light source, an image sensor, an authentication unit, and a template storage device (for example, a smart card or a database repository). The authentication unit has four key components:

1. Video input/output (I/O) device for capturing data forwarded from the camera sensors.
2. An LED power controller to manage the array of LEDs.
3. A central processing unit (CPU) core for processing.
4. An I/O controller that performs a variety of tasks, such as deactivating door locks for physical access applications.

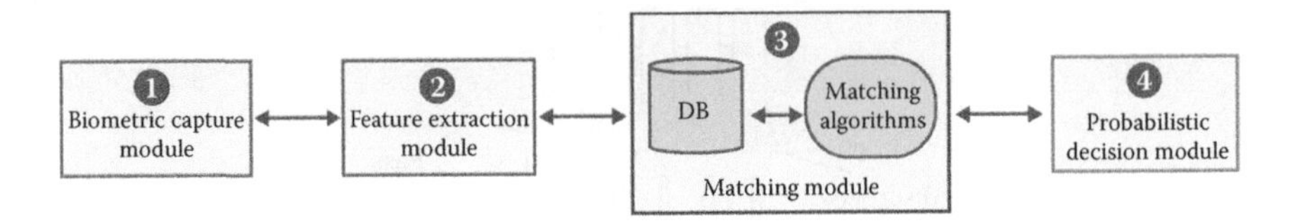

FIGURE 4.3 Components of a VPR system.

As shown in Figure 4.3, the VPR system performs four key tasks: (1) capturing the vein pattern image, (2) extracting the vein pattern features from the image, (3) providing a matching score, and (4) rendering a decision. Although each VPR system may differ in the specific ways that it accomplishes these tasks, each generally follows a sequence of actions similar to the following:

In Task 1, the vein pattern image data is acquired and placed in the memory of the CPU, which usually adjusts the brightness of the light scanners via the LED power controller to account for any externalities (such as bright ambient light). In Task 2, the vein image is normalized to account for any positioning differences that the hand or finger presents. The VPR system basically detects the outline of the finger or hand, and then rotates the image to the appropriate alignment. The VPR system performs an extraction of the distinctive feature patterns. In Task 3, the VPR system performs template comparisons. It correlates the similarity of the extracted feature pattern (e.g., the sample template) with the registered pattern (e.g., the enrollment template), and assigns a correlation value (e.g., probability of a match). Finally, in Task 4, it evaluates that correlation value and provides an authentication decision—match or no match.

Vein pattern recognition offers an array of advantages over other biometric techniques, including the following:

- *Ease of feature extraction*—Veins provide large, robust, relatively stable and distinctive (to the CCD camera) biometric features. With the right equipment, vein patterns are easily acquired and compared through the use of near-infrared light, which has no injurious effects on the human body.
- *Spoofing resistant*—Vein patterns are not easily observed (with the naked eye), damaged, obscured, or changed. They do not exhibit latency properties, and they are quite difficult to replicate. This makes them highly resistant to spoofing. Moreover, all the VPR systems digitize and encrypt the data in the templates.

- *High accuracy*—Vein patterns have very low false acceptance levels (<.0001%) and low false reject levels (<.01%). Further, less than 0.2% of most user populations are unable to enroll. Vein pattern biometrics can differentiate between identical twins; that claim is shared only by ocular recognition (iris and retina), fingerprint, and facial thermography modalities. Moisture, precipitation, minor dirt, and cuts do not impact the accuracy of VPR systems.
- *Environmental freedom*—Vein patterns are much less susceptible to many external factors in the way that fingerprinting, facial recognition, or iris scanning can be. For example, indoor lighting has a limited effect on vein authentication. With appropriate screening, some VPR systems can achieve authentication under direct sunlight as well, rendering authentication possible whether the individual is indoors or outdoors.
- *User friendliness*—Vein pattern technology is considered to be very privacy enhancing. It is fast (<2 seconds), user intuitive regarding the placement of one's finger or hand, hygienic, and generally bereft of cultural bias and criminal stigmas. Additionally, automated audio and visual guidance during both enrollment and verification transactions provide helpful feedback.
- *Small footprint*—Sensors are quite compact and are now approaching the small size of the fingerprint readers; and they are continuing to rapidly downsize, increasing portability for use with mobile devices. The Fujitsu PalmSecure™ palm vein sensor, shown in Figure 4.4, is only 35 × 35 × 27 mm (1.38 × 1.38 × 1.06 in.), weighs

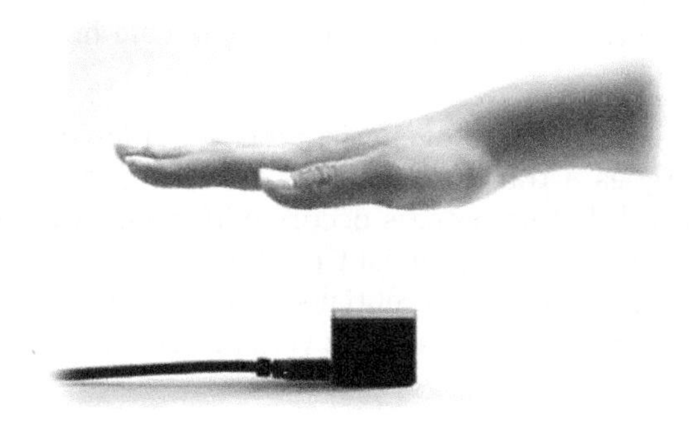

FIGURE 4.4 Palm vein authentication device. (Courtesy of Fujitsu Ltd.)

FIGURE 4.5 Logical access reader. (Courtesy of Hitachi Ltd.)

only 50 grams (1.76 oz.), and with its small footprint it can easily be carried in a laptop bag. The Hitachi H1 Logical Access Finger Vein reader, shown in Figure 4.5, is not much larger at 59 × 82 × 74 mm (2.32 × 3.23 × 2.91 in.), and it weighs only 96 grams (3.4 oz.). These devices are continuing to rapidly downsize, increasing portability for use with mobile devices.

In Chapter 2, we discussed what makes a good biometric. We named and described seven evaluation criteria: (1) uniqueness, (2) permanence, (3) universality, (4) collectability, (5) acceptability, (6) performance, and (7) resistance to circumvention. In the following, we will apply these seven evaluation criteria to VPR systems to evaluate how well the VPR modality stacks up.

1. *Uniqueness*—During ontogenesis,* the vascular network pattern undergoes a transformation as the arteriovenous network is formed. While this process occurs within genetic constraints, it enables statistically significant random pattern variation such that individuals can be differentiated based on the distinctiveness of those features. This means that vein patterns in one's finger or

* Ontogenesis describes the origin and the development of an organism from the fertilized egg to its mature form.

one's hand are highly unique to each individual, providing the basis for personal verification. As has been mentioned, the patterns differ from one hand to the other and from finger to finger. The exceptional uniqueness associated with vein pattern recognition leads to very low false acceptance rates.

2. *Permanence*—The basic pattern of blood vessels is formed during the fetal stage.* Blood vessels maintain a relatively stable vascular structure due to tight interactions among the cells that compose them. Generally, fingers and hands have a consistent flow of blood. A growing body of empirical evidence supports the widely held belief that vein patterns generally remain invariant with age. However, in the case of young children, more analysis is required to determine to what degree, if any, their vein patterns change. Some blood vessels may incur some blocking with disease, and new blood vessels may form.† However, both possibilities are quite rare.
3. *Universality*—Every human has an extensive pattern of veins and arteries that are essential for circulating oxygen and nutrients to the body's extremities. The thick veins just under the skin surface of the hands and fingers exist in all people (unless they have suffered congenital or traumatic amputations). Veins tend to be 0.3 to 1.0 mm in thickness, making them readily assessable for quantitative measurements. As mentioned earlier in this chapter, one's hands and fingers are not the only areas that could be used for biometric vein pattern recognition; however, they are the most convenient.

A key advantage of VPR systems is that they can be used by over 99.8% of the population. They can be comfortably used by a broad segment of any population, as underscored by its very low failure to enroll (FTE) rate. Additionally, VPR systems do not appear to be hampered by temperature variation within normal temperature ranges common to human habitation. The high usability of VPR systems increases the overall security, since it is primarily the individuals who cannot use a particular biometric technique that must use a weaker or less efficient authentication method.

* Fetal stage generally runs from the end of the embryonic stage until birth.

† Angiogenesis is the formation of new blood vessels, generally occurring between a tumor and surrounding tissue so that the tumor can be nourished.

Although VPR systems tend to be highly usable in most environments, there are several situations that can limit that usability. The following are known VPR limitations:

- *Presence of carbon or ink*—The presence of carbon or various inks on the outer surface of a finger or hand could block the camera sensor's ability to make a quality infrared (IR) image. Carbon absorbs light of all wavelengths, including near infrared. Thus, this technology may not work well in venues such as coal mines where users are covered with carbon soot. Tattooing ink has a similar obfuscating affect. Although tattooing might occur on the dorsal side of the hand or finger, such situations are rare, limited only to special groups.
- *Too much light flooding the VPR sensor*—There are individuals whose hands and fingers are so small in comparison with the rest of the population that their use of vein pattern readers enables excessive light bleed to occur. The excessive light blurs the image that the sensor is trying to capture. Most VPR systems use a fixed camera and a much smaller than anticipated finger or hand might not cover the light well enough. This issue can be overcome by controlling the light intensity of the LEDs or the sensitivity of the camera, or perhaps refocusing the camera on a smaller surface area, as might be done for readers designed for children.

4. *Collectability*—Performing a VPR transaction is simple, nonintrusive, and very fast. Placing one's finger or hand on the reading device is convenient and intuitively easy, and training people to use VPR systems can occur very quickly. Generally, using VPR systems does not cause people concern nor increase their anxieties. Moreover, extracting and measuring the vein pattern data is straightforward. Scanning a vein pattern is not affected by aging or small cuts, nor do scars or skin color affect the image outcome. Additionally, the moisture on or roughness of one's skin has no effect on the performance of VPR systems. Generally, the only external factor to significantly impact the authentication process is intense bright light, which can be controlled in most situations. Most VPR systems are able to collect an individual's vein pattern data, make a comparison, and return a decision in less than 2 seconds. It is that fast and that easy.
5. *Acceptability*—Vein pattern recognition has already been implemented with great success in Asian markets where there is strong

resistance to fingerprinting due to its association to criminal activity. On the other hand, Asians eagerly accepted VPR systems because they are hygienic (near contactless), convenient, and highly secure. Most of the VPR vendors refer to their modality as "contactless." They mean that there is no direct contact with the camera sensor as there is with fingerprint and hand geometry modalities. However, there is usually some minor contact involved. The palm and back of the hand vendors often use a stabilizing support to rest the hand; and finger vein vendors tend to use a small button that is pressed by the tip of the finger, or the fingertip fits into a special groove to position it the same way for each sensor reading. These methods enable a proper alignment of the hand or finger. However, all VPR systems position their scanning cameras several centimeters away from fingers or hands, hence the contactless nature of the read.

In many Western countries, the privacy protection characteristics of VPR are already gaining a substantial following. That is, since the veins exist within the body, a person's vein pattern cannot be readily viewed without special equipment, and thus, the pattern cannot be easily forged. The use of vein pattern recognition is not difficult, inconvenient, or intrusive, nor is there any negative cultural context associated with this biometric modality. Thus, VPR technology is becoming one of the most acceptable biometrics to user populations.

6. *Performance*—Performance is effectively an umbrella term that describes the tradeoff of accuracy and speed. For example, both accuracy and speed are somewhat affected by the size of the template. The general template size (e.g., 300 bytes to approximately 2 Kbytes) of VPR systems is detailed enough to offer significant accuracy, yet small enough that an efficient algorithm can quickly produce results. It is a common misconception that the larger the template size, the higher is the level of accuracy. Actually, the accuracy of an authentication device is primarily based on the clarity of the captured data images and the strength of the algorithm used in the authentication process. To underscore this point, let us compare two-dimensional (2D) face recognition with retina scanning. Face recognition, which uses relatively large templates of 2 to 3 Kbytes, is known to have limited accuracy. Retina scanning ranks among the most accurate modalities, yet it produces compact templates of approximately 200 bytes. However, to meaningfully

assess accuracy, each company's data gathering methods need to be standardized to achieve an accurate benchmark.

VPR systems are quite accurate. Many VPR systems are among the most accurate biometric systems, but that is a function of the biometric implementation as well as the modality. For example, in an effort to reduce prices of biometric scanners, implementations of some systems (of various modalities) are rendered less accurate than they might otherwise be. These less accurate systems may use less robust algorithms, employ a less accurate sensor, limit the number of templates compared, or reduce the processing efforts to save time or cut costs.

7. *Resistance to circumvention*—As Figure 4.6 illustrates, vein pattern recognition is a very difficult biometric modality to circumvent. If vein pattern recognition technology is competently deployed, it is very difficult to spoof. Unlike facial recognition or speaker recognition, which can be imaged or recorded, respectively, with conventional equipment and without the individual's knowledge or consent, vein imaging requires special equipment and special technology. Vein patterns are not susceptible to tampering, and, unlike fingerprints, vein patterns are difficult to acquire in daily life. The absorption of near-infrared light by the hemoglobin flowing through the veins is a built-in liveness detector, in addition to an array of manufactured liveness detectors in most VPR systems. This increases the difficulty in spoofing VPR systems.

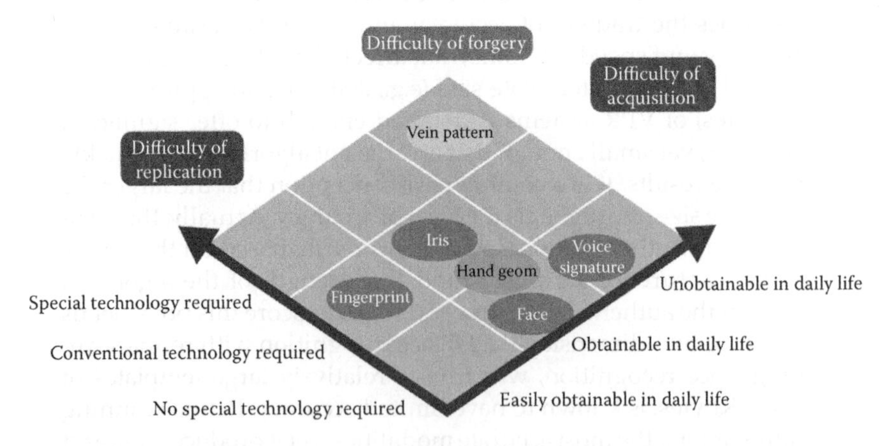

FIGURE 4.6 Difficulty in spoofing VPR systems. (Courtesy of Hitachi Ltd.)

PEOPLE CONSIDERATIONS

In designing any biometric system, one must consider people's attitudes and perceptions. The growth of a given technology is highly dependent on how comfortable end users become in using it. A superior biometric system is one that widens usage, reduces user anxiety, and lessens any preexisting user reluctance toward its use. Therefore, it is crucial that an enterprise consider the human factors when selecting and deploying a biometric system. Vein pattern recognition technology was developed to provide a highly desirable, fast biometric technique as well as support people considerations such as hygiene, privacy, convenience, and personal security. Vein pattern recognition offers an array of "people friendliness" advantages vis-à-vis other biometric technologies.

- *Near contactless*—As previously stated, there is no direct contact with the surface of the camera's sensor, and there is minimal physical contact with the vein pattern readers. One's finger or hand merely aligns with the reader so that the CCD camera can acquire a consistent vein pattern image. Some VPR systems do not even require specific placement of the hand; a slow wave over the sensor is all that may be needed. The use of the near-infrared light enables nonintrusive imaging that ensures ease and cleanliness for the user experience. There is no pressing of the finger or hand against a sensor, causing it to get dirty or oily, thus decreasing the success of the authentication. Moreover, such minimal physical contact limits the spread of bacteria and viruses from user to user.
- *Privacy protection*—A key advantage of vein pattern recognition technology is that it can support access authentication in a privacy-enhancing way. One's finger vein pattern cannot be read without the knowledge and consent of the individual. There are no latent properties with VPR technology; that is, one cannot leave behind a VPR pattern as one can leave DNA and fingerprints. What this means to privacy advocates is they now have available to them a privacy-oriented biometric. And for ultimate privacy and security, one's vein recognition template can be placed on a smart card (each template requires .3 to 2 Kbytes in programmable memory, depending on the VPR submodality and the system implementation), a laptop, or a personal digital assistant (PDA). In those scenarios, the template is safeguarded in a token or device that the individual controls, and the biometric

stays in one's finger or hand where it belongs. That is truly privacy enhancing.

- *Convenience*—There is minimal user effort in using a VPR system whether one is using the back of the hand, the palm, or the finger. Anyone who can place a hand or finger on a vein pattern reader can use this biometric system. A VPR reader requires minimal positioning of the hand or finger, and it takes less than 2 seconds for the reader device to read the vein pattern, process the template, and perform a match algorithm. It is a very natural, easy process. There is also minimal user training required, and there are no special requirements. Finally, the false rejection rate is extremely low, thus not delaying enrolled users with denied access and a requirement to re-verify.
- *Highly secure*—VPR biometrics are closed view in that the physical traits to be measured are internal to the human body and are hidden from common sight, rendering it far more difficult to inappropriately capture and replicate. Open-view biometrics such as fingerprints, voice scanning, and facial scans can be more easily recorded or photographed, and are more susceptible to covert capture and use by fraudsters to spoof control systems.

LIGHT IMAGING

Vein pattern technology uses near-infrared light, typically wavelengths within the range of 700 to 1200 nm, to visualize and identify unique vein patterns. As near-infrared light projects deeper into living tissue, it can be rapidly diffused, and the contrast of veins to their background can quickly deteriorate. Dr. Mitsutoshi Himaga uses the analogy of placing a swizzle stick in a glass of milk.* The swizzle stick can be readily seen from outside the glass when it is in close proximity to the edge of the glass. However, the swizzle stick becomes gradually invisible when it is moved further into the center of the glass of milk. This is due to the principle of light diffusion.†

* Himaga, of Japan-based Hitachi-Omron Terminal Solutions, is a well-known practitioner of vein pattern recognition technology.

† Light diffusion is defined as the scattering of light by reflection from a surface or transmission through a translucent substance.

Infrared light-emitting diodes (IR LEDs) are the most commonly used light source for vein pattern imaging. IR LEDs have a rich history of use in household appliances such as TV remote controllers. The arrangement of the light source differs according to its use. Finger vein imaging systems typically require small and oblong fields of view, and therefore use more linear arrays of IR LEDs. Some palm or back of the hand VPR systems do require a gridlike or circular light source arrangement.

As was discussed in Chapter 3, it is exceedingly difficult to produce a biometric image that is pixelwise identical to the enrollment image due to an array of environmental and user variables. Therefore, many VPR systems continuously capture the presented sample with multiple illumination variations. Each of the sample patterns is matched to the enrollment template one by one in real time; and the VPR system continues this iteration until the presented sample is either accepted or rejected. Therefore, some VPR vendors strive to create an illumination control algorithm that produces optimized images as quickly as possible so that an access attempt can be processed and adjudicated without attenuation.

There are two conventional approaches to illuminate the vein pattern of a subject: (1) diffused illumination to reflect the target features, also referred to as the light reflection method, or (2) direct illumination such that the image is based on transparency (one can actually "see" through the skin to the veins), also referred to as the light penetration method or transillumination. Light reflection captures images of blood vessels even through thick skin. The light source and image sensor are placed on the same side of the hand. As the light source illuminates the body tissue, it is reflected back, and the CCD sensor uses the light reflected off the skin's surface to reveal an image of the blood vessel pattern. Hand vein recognition (both for palm and back of the hand) tends to use light reflection. The light reflection method does offer some design advantages. Since the light source and image sensor can be close together, the scanner device can be more compact.

With direct illumination, the body part is placed between the image sensor and the light source, and the near-infrared light passes through the surface of the skin. The shadow cast by the hemoglobin absorption is detected, and the vein pattern can be seen by the IR camera. This enables capture of high-contrast and highly detailed vein pattern images. However, light can only penetrate areas of the body where the skin is not too thick. A finger, for instance, is an ideal body part for light transmission via near-infrared light. Moreover, extensive testing

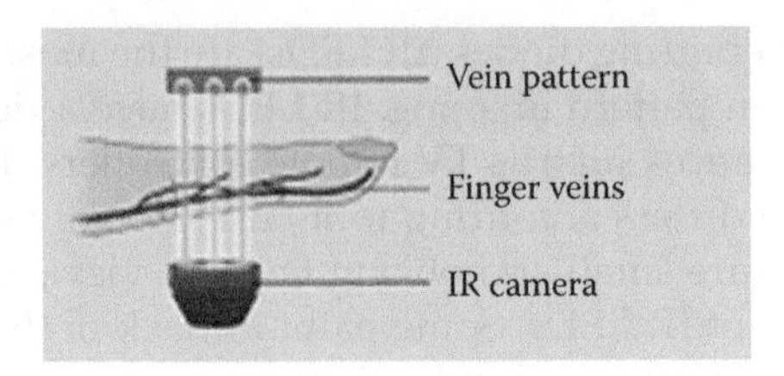

FIGURE 4.7 Top-lighting system. (Courtesy of Hitachi Ltd.)

has confirmed that direct illumination leaves no harmful effects on the human body.

There are three configurations of the direct illumination method that deserve a quick review: top-lighting systems, side-by-side lighting, and bottom lighting.

Top lighting places the infrared light source on the opposite side from the camera. As illustrated in Figure 4.7, near-infrared rays penetrate the finger to enable the camera sensor on the opposite side to visualize the vein patterns that are illuminated. This is the earliest and most straightforward implementation method. It has been used for a variety of finger vein reader models, as shown in Figure 4.8 (physical access) and Figure 4.9 (logical access).

The control of illumination is relatively straightforward and easy with top lighting. The housing for the light source protects the camera from unwanted ambient light and dust accumulation, which might otherwise

FIGURE 4.8 Secure vein attestor. (Courtesy of Hitachi Information & Control Solutions Ltd.)

FIGURE 4.9 Logical access finger vein reader. (Courtesy of Hitachi Ltd.)

deteriorate the image quality. Top lighting is considered a robust solution regarding environmental changes. However, the light-source housing may yield a spatial "footprint" that is too large for some portable devices such as PDAs or mobile phones. Such systems also present a challenge to cleaning the surface of the sensor once it becomes dirty. Moreover, depending on the actual reader design, some people feel quite anxious about inserting their fingers into the "hole-like" housing, including concerns regarding spiders and general fear of the unknown.

Another implementation was demonstrated at the 2007 Tokyo Motor Show, where the driver presents his finger to the reader embedded on a steering wheel to start the car's engine (Figure 4.10).

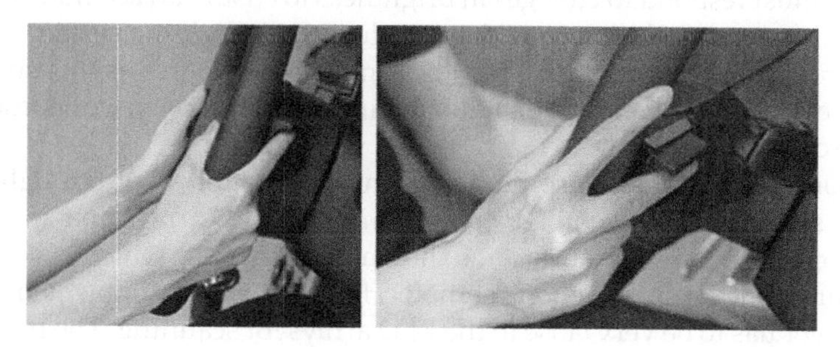

FIGURE 4.10 Finger vein reader embedded in steering wheel (prototype). (Courtesy of Hitachi Ltd.)

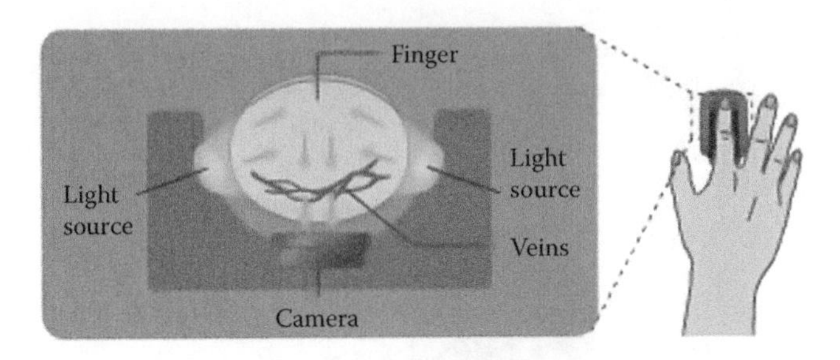

FIGURE 4.11 Cross-section view of finger. (Courtesy of Hitachi Ltd.)

Some implementations use side lighting, whereby light sources are positioned on both sides of the finger, and the sensor captures the vein pattern image. The infrared rays emitted by the light source propagate inside the finger and some of them enable imaging by the camera placed beneath the finger (see Figure 4.11). This method uses adaptive light control to image clear vein patterns; and it enables the high-contrast imaging while enabling an open, ceiling-less scanner device. This implementation supports medium-sized enclosures, has a user-friendly design with very low psychological barriers, and is easy to maintain. Unlike some top-lighting systems, the presented finger is not hidden from view. The user simply places his or her finger on the device surface for scanning. Because two separate LED banks are illuminating the finger, the illumination control required to obtain homogeneous contrast throughout the field of interest is relatively complicated. Moreover, this method usually requires more CPU power. Finally, because of the open-air optical design, it may be more sensitive to environmental light. Due to variations in finger size, particularly the width, the brightness of the light source must adapt to and the image sensor must respond to changes in brightness to ensure an accurate image. Some finger vein solution providers produce a side-lighting system with a semitransparent cover (which may be removable indoors), as this enables the user to see inside the cover, reducing individual anxiety, while restricting excessive light.

Some VPR systems use the third implementation—bottom lighting. These systems usually embed a pair of infrared LED arrays and an infrared camera on the same surface, as shown in Figure 4.12, which looks very similar to the light reflection method. However, in this implementation the finger has to be very close to the LED arrays for scanning. The two sets of LED arrays project into the finger to enable the camera to visualize the vein patterns in the same manner as side-lighting implementations. This

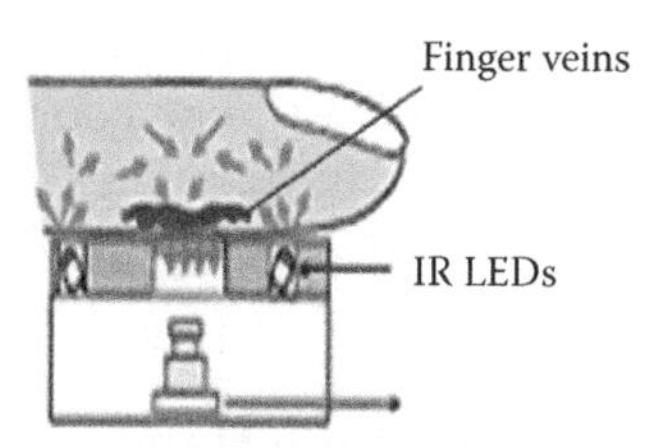

FIGURE 4.12 Illustrated bottom-lighting system. (Courtesy of Hitachi Ltd.)

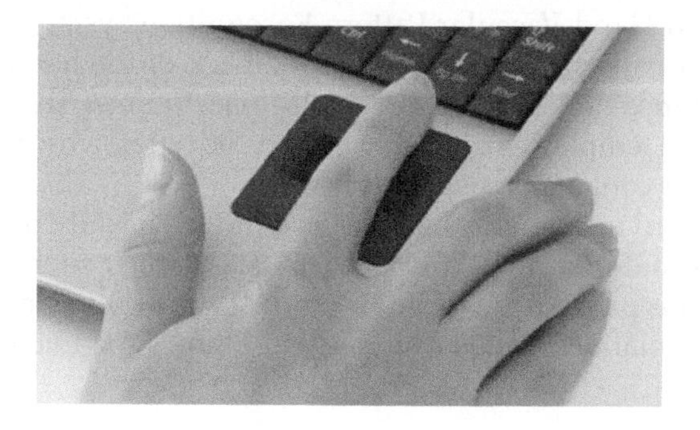

FIGURE 4.13 Bottom-lighting system in a laptop. (Courtesy of Hitachi Ltd.)

type of imaging is inexpensive, enables a very small form factor, is easy to maintain, and has no psychological barriers to use. On the other hand, the illumination control complexities are similar to side lighting: It is sensitive to excessive ambient light, and it can be difficult to always position the finger in the same manner because there are limited guides. Figure 4.13 shows a reader unit that employs bottom lighting.

As this discussion of illumination has suggested, the most suitable imaging method and implementation is highly dependent on the application. Additionally, the performance of the various implementations can best be improved by suppressing the influence of external factors.

VPR SUBMODALITIES

The VPR technologies map vein patterns from different locations of the human body. The currently available VPR modalities obtain their images from one's hand: the finger (both palm side and back side), the hand palm, and the back of the hand. Future VPR technologies *might* use different parts of the human body. Theoretically, one could map vein patterns of the face, wrist, legs, feet, and toes. The following sections discuss the current commercially available submodalities and their vendor systems in more detail.

Back of the Hand

Techsphere Company Ltd., based in Seoul, Republic of Korea, has developed a vein pattern scanner, the VP-II™ reader, which it introduced in 2001. The VP-II Hand Vascular Pattern Recognition System verifies users based on vein patterns extracted from the back of the hand. It purports a false acceptance rate of .0001% or a false reject rate of .01%. It produces fully encrypted templates of just fewer than 300 bytes, which obviates any storage issue or congestion over a network.

The standard pose for the back of the hand is typically positioning the hand with the outer or dorsal side toward the scanner such that the camera's direction is parallel to the *z*-axis on the back of the hand. Figure 4.14 depicts an orientation of the left hand that follows the Euclidean coordinate system.

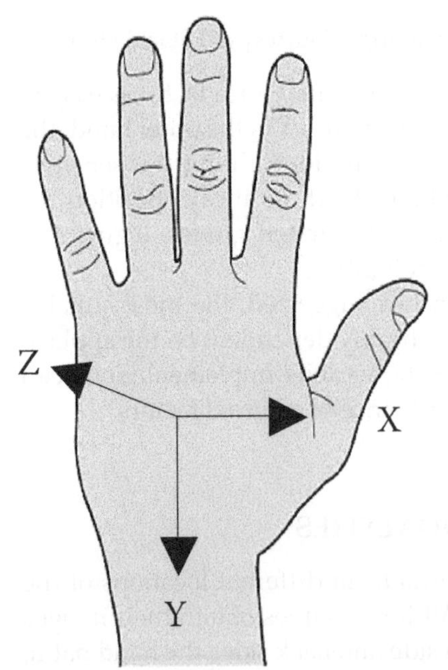

FIGURE 4.14 Orientation of the human hand, palm-side down.

Identica Corporation

The VP-II is exclusively marketed by U.S.-based Identica Holdings Corp. in North, Central, and South America; Europe (extending from the United Kingdom to the whole of the Russian Federation); and Israel. In conjunction with the VP-II, Identica integrates its proprietary universal controllers and application software, including smart card and proximity card interfaces. To provide for adverse environmental conditions such as extreme heat, cold, and pollutants, Identica shields its identification devices with its WeatherShield™. Identica markets its products through resellers, integrators, and if required, directly to large, government and institutional end-use customers.

The VPR scanner uses near-infrared technology to capture an individual's unique vein pattern from below the surface of the skin on the back of the hand. This simple-to-use, fast, hygienic, and highly accurate solution enables a unique personal template to be captured, encrypted, and then stored a variety of ways. It verifies users through implementation of its patented recognition algorithm.

The VP-II system can be configured to operate in stand-alone or networked environments using proprietary, TCP/IP (transmission control protocol/Internet protocol) and Wiegand™ protocols. The VP-II product line consists of the VP-II X scanner (see Figure 4.15), a physical access control system that can be configured to work with the integrated Universal Controller, and Identica's IONcontrol-X software to support a fully integrated network of VP-IIs. The VP-II X supports HID* and MIFARE™† smart cards, and is fully compatible with various U.S. government credentials including TWIC,‡ Personal Identity Verification (PIV), and CAC.§

Using Identica's IONcontrol Software Development Kits, systems integrators and other solution providers can rapidly develop fully integrated

* HID Corp., based in Irvine, California, produces proximity cards and readers and as well as its iClass smart card and readers, and is recognized globally as a de facto standard for physical access control.

† MIFARE is arguably the most widely installed contactless smart card technology in the world with over 500 million smart card chips and 5 million reader modules sold. It is a trademark of NXP Semiconductor, a spin-off company of Philips Semiconductors.

‡ TWIC refers to Transportation Worker Identification Credential. It is a common identification credential for all personnel requiring unescorted access to secure areas of regulated facilities and vessels, and all mariners holding Coast Guard-issued credentials.

§ CAC is an acronym for Common Access Card, a U.S. Department of Defense (DOD) smart card issued as standard identification for military personnel and defense contractors.

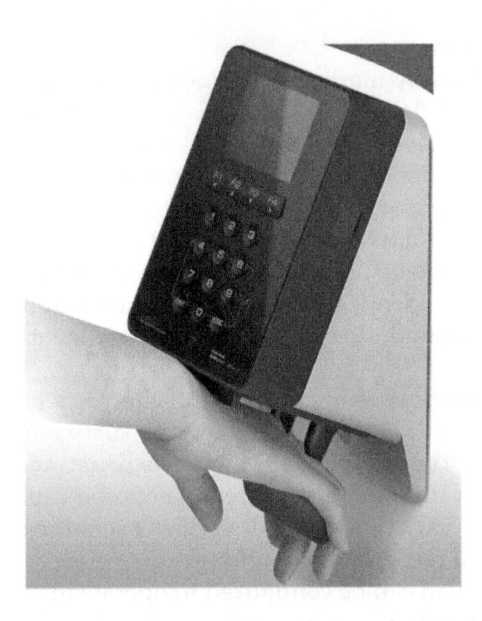

FIGURE 4.15 VP-II X hand scanner. (Courtesy of Identica Corp.)

applications with current, legacy, and future TCP/IP-based entry management, time and attendance, and credential verification systems.

A user simply inputs a PIN (shown in Figure 4.16) or waves a contactless smart card or proximity card, and then places his hand to the Identica VP-II scanner, which employs patented recognition algorithms to capture and encrypt the unique vein patterns on the back of one's hand when it is placed on the guide beneath the reader (as shown in Figure 4.17).

With the option of storing the encrypted unique personal template on media such as a smart card, the scanner communicates with Identica's proprietary integrated Universal Controller and verifies the user against the stored template on the media. Therefore, the stored template stays with the user, thus eliminating any lingering privacy concerns.

The VP-II integrated Universal Controller can be employed using TCP/IP to communicate with the network to store and access the biometric template. The user either inputs a PIN or waves a contactless smart card or proximity card, and then presents the back of his hand for identification. The sample vein pattern template is then compared to the

FIGURE 4.16 Using PIN pad on VP-II scanner. (Courtesy of Identica Corp.)

FIGURE 4.17 Placing hand on the VP-II scanner. (Courtesy of Identica Corp.)

network-stored reference template, and the identity verification message is sent to the respective application.

The VP-II system can be configured to log in and log out users in a stand-alone or networked environment. Logging and report generation are accomplished by an intuitive and easy to use graphical user interface (GUI).

FIGURE 4.18 Outdoor mount for VP-II scanner. (Courtesy of Identica Corp.)

Identica's VPR scanners are treated with antimicrobial silver to combat biofilm.* The antimicrobial silver attacks the microorganisms at the cellular level. Identica asserts that by applying antimicrobial silver to the external components of their VP-II scanners, 99.9% of the biofilm are destroyed. This further underscores Identica's focus on hygiene.

Identica targets its VPR solutions to access control, and time and attendance applications, but it can be used at any organization that needs to have an undeniable audit trail of when and where a known individual enters a building or begins a specific task. Identica's target markets are government, transportation (e.g., airports), financial institutions, gaming industry (e.g., casinos), health care, and utilities.

Current high-profile users of the back of the hand VPR systems include the South Korean government, Inchon International Airport, the Tokyo Police Data Center, SunFirst Bank (based in St. George, Utah), the airport staff in Seoul, the Toronto International Airport, the University of Ottawa, and port workers at the Port of Halifax. The VP-II demonstrates its ruggedness in Figure 4.18, an outdoor mount. Identica's WeatherShield, an automated heated outdoor enclosure, ensures consistent outdoor operation in harsh climates and extreme temperatures. Identica and Techsphere target government agencies, large enterprises, luxury apartments, hospitals, dormitories, universities and schools, data centers, and warehouses for the applications shown in Table 4.1.

* Biofilm is a thin layer of microorganisms such as bacteria or fungus, which form on and coat various surfaces, held to those surfaces by the material the microorganisms produce.

Table 4.1 Targeted Industries

Building physical access	Used in place of keys or keycards to gain access to secure areas of buildings
Immigration	Reduce unlawful entries, speed access for legitimate travelers
Airport check-in	Speed check-in procedure
ATM access	Access bank/ATM accounts without card; no need to remember PIN
PC/LAN logon	Eliminates the need for logon IDs and passwords for workstations
Retail payment	Reduce losses from bad checks and credit card fraud
E-commerce	Verify identity for online purchases and payments, as trust in e-commerce transactions is essential
Government agencies	Reduce fraudulent activities related to identity theft, e.g., social insurance and welfare claims

Source: Courtesy of Identica Corp. and Techsphere Co. Ltd.

The Palm

The palm is another convenient area to employ vein pattern technology. Palm vein images are taken with the palm area basically flat, not bent, and each finger boundary exposed to the camera, with fingers straight. Figure 4.19 depicts an orientation of the left hand that follows the Euclidean coordinate system.

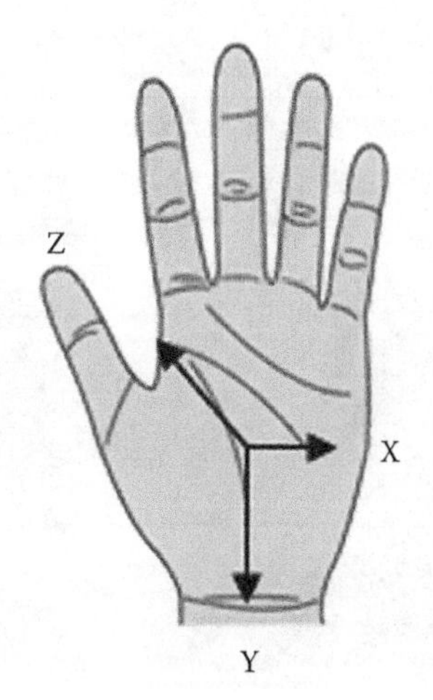

FIGURE 4.19 Orientation of the human hand, palm-side up.

Fujitsu Ltd.

Fujitsu's palm vein authentication technology consists of a small palm vein scanner that is natural to use, fast, and highly accurate. A user places his palm a few centimeters over the scanner and in less than a second it reads his unique vein pattern. A vein picture is taken and the pattern is registered. Through the testing of 140,000 palms from 70,000 individuals, Fujitsu determined that it could achieve a high accuracy of a 0.01% false rejection rate or a 0.00008 false acceptance rate. That accuracy rate was confirmed in testing with the International Biometric Group (see Chapter 6). Palm vein authentication uses the vein patterns of an individual's palm as personal identification data. A palm has a broad, complicated vein pattern and thus contains a wealth of differentiating features for personal identification. Figure 4.20 clearly depicts how the palm vein pattern is visualized by the CCD camera in the palm vein reader. The hemoglobin coursing through the palm's veins absorb light having a wavelength of about 760 nm within the near-infrared area. Figure 4.21a shows a human hand hovering over a palm vein scanner. When the infrared image is captured, only the vein pattern is visible as a series of dark lines, as shown in Figure 4.21b. Based on this feature, the vein authentication device translates the black lines of the infrared image as the vein pattern of the palm

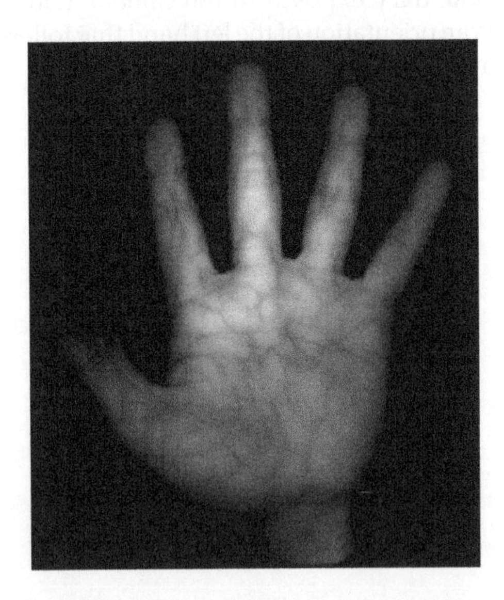

FIGURE 4.20 Palm vein pattern. (Courtesy of Fujitsu Ltd.)

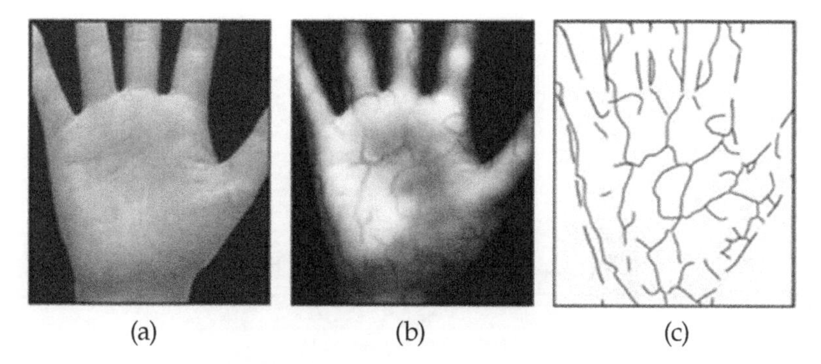

(a) (b) (c)

FIGURE 4.21 Near-infrared image. (Courtesy of Fujitsu Ltd.)

as depicted in Figure 4.21c, creates a sample template, and then matches it with the previously registered vein pattern template of the individual.

The contactless palm vein authentication technology consists of image sensing and system software. The palm vein sensor captures the image of the user's vein pattern by bathing it in near-infrared light. The near-infrared lighting is controlled depending on the illumination around the image sensor, and it captures the palm image regardless of the position and movement of the palm. The software algorithm then matches the translated vein pattern with the enrollment pattern while measuring the position and orientation of the palm by a pattern-matching method. Implementation of a contactless verification system such as Fujitsu's PalmSecure system enables applications in public places or in environments where hygiene standards are required, such as in medical applications (see Figure 4.22).

The opportunities to implement palm vein technology span a wide range of vertical markets, including security, banking, health care, commercial enterprises, and educational facilities. Commercial applications include physical admission into secured areas; log-in to PCs or server systems; access to points-of-sale (POS), ATMs, or kiosks (see Figure 4.23); positive ID control; and other industry-specific applications. With the new palm vein authentication device and with considerable experience in image recognition, Fujitsu has become a leader in providing solutions for the biometric security industry.

Fujitsu created the PalmSecure scanner to meet the needs of its customers, many of whom associated fingerprint scans and face recognition methods with the police on a psychological level. Other customers had expressed sanitary concerns regarding touching what everyone else has touched.

FIGURE 4.22 Contactless palm vein recognition kiosk. (Courtesy of Fujitsu Ltd.)

FIGURE 4.23 Free-standing contactless palm vein recognition kiosk. (Courtesy of Fujitsu Ltd.)

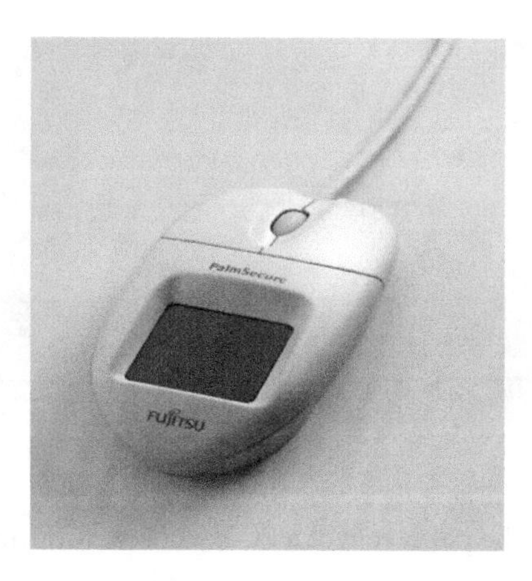

FIGURE 4.24 Palm vein mouse. (Courtesy of Fujitsu Ltd.)

Fujitsu is targeting corporate desktop computer access with its new scanner and PC mouse (as seen in Figure 4.24). Fujitsu Laboratories Ltd. has developed the palm vein image sensor on the mouse surface to read the palm resting on its top.* A built-in camera captures an image of the palm's subcutaneous veins and verifies the individual. According to the experiments conducted by Fujitsu Laboratories, the palm vein patterns of about 700 people were stored in a database system, and all were subsequently identified correctly. Fujitsu believes that the biometrics computer mouse can achieve an equal error rate of 0.5% or less.

One way that Fujitsu continues to improve upon this promising new technology is by shrinking the scanner. As depicted in Figure 4.25, palm vein authentication has been integrated into a PC keyboard. As a future goal, Fujitsu is enthusiastically working to incorporate the scanner in mobile phones. Fujitsu's success in such a venture would drastically change the way mobile phones are used. The security of handheld electronic devices will become increasingly important as they store more personal information. In all these applications, the key to securing assets and data would be in the palm of one's hand.

Japan had seen a rapidly growing problem in the illegal withdrawal of bank funds using stolen or skimmed fake bank cards. To address this, palm vein authentication has been used for customer confirmation of transactions at bank windows or ATMs. The smart card from the customer's

* Fujitsu Laboratories Ltd. is a premier center for basic and applied research, and it is a wholly owned subsidiary of Fujitsu Ltd.

FIGURE 4.25 Using the palm vein at a PC keyboard. (Courtesy of Fujitsu Ltd.)

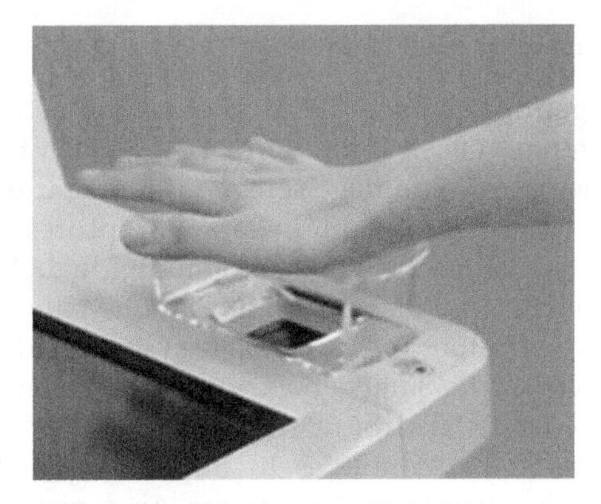

FIGURE 4.26 Using the palm vein at an ATM. (Courtesy of Fujitsu Ltd.)

bank account contains the customer's palm vein pattern and the matching software of the palm vein patterns. As shown in Figure 4.26, with Fujitsu's VPR system, a user inserts his smart card into an ATM, and then is prompted to hold his hand (palm down) near an infrared light source. A palm vein authentication device at the ATM scans the customer's palm vein pattern and transfers the sample template into the smart card. The customer's palm vein pattern is then matched with the enrollment template in the smart card. Since the enrollment template is not released from the smart card, the security of the customer's vein pattern is preserved.

To broaden the applications that palm vein technology can address, and to further improve the convenient use of palm vein pattern

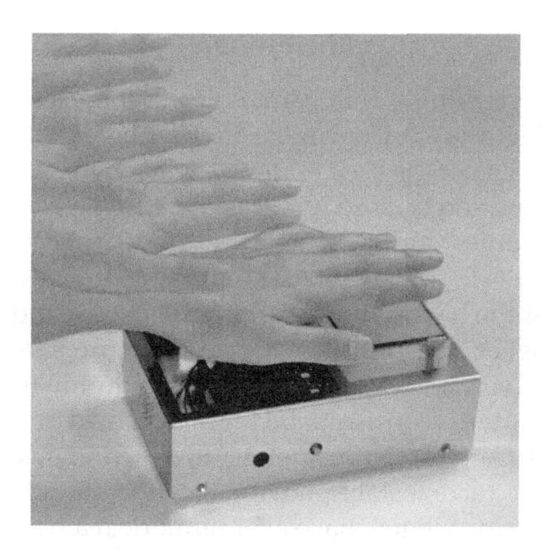

FIGURE 4.27 Palm vein high-speed image capture prototype. (Courtesy of Fujitsu Ltd.)

recognition, Fujitsu Laboratories announced in April 2009 that its newest high-speed capture technology can now operate while one's palm is in motion. Figure 4.27 shows the high-speed image capture prototype, which requires only 1 millisecond to capture the vein pattern image and authenticate the individual with the same accuracy Fujitsu previously offered in earlier palm vein technology iterations. Moreover, it achieves a frame rate of 30 frames per second.

To support a 1 millisecond exposure with sufficient clarity for an accurate image capture, Fujitsu Laboratories optimized the lighting controls and the optical structure of its reader. Thus, even if the palm is moving rapidly at the average speed of walking, about 1 meter per second, the sensor will acquire sharp vein pattern images without blurring. To maintain its high authentication accuracy, Fujitsu developed a function that enables automatic selection of the best image for authentication from the stream of images captured by its high-speed image capture module. This high-speed imaging technique will enable Fujitsu to authenticate identities with the same ease and speed of systems that use contactless cards (both smart cards and proximity cards) over turnstiles or doors for physical access applications. Thus, Fujitsu's new palm vein system can expand into additional high security venues where speed of contactless authentication is a key criterion.

The Finger

The standard pose for finger vein pattern recognition is the finger held out straight, unbent (Figure 4.28). Depending on the submodality, the front or ventral side of each finger, or the back or dorsal side can be used. Of course, any finger on either hand can be used with a vein pattern recognition system.

Several companies support finger vein authentication. Finger vein systems and hand vein systems have similar advantages and limitations. Some believe that the back of hands and palms house more complex vein patterns than fingers, and thus provide more features for pattern matching. Finger vein advocates counter that the transillumination of the fingers renders its vein pattern features to be highly distinct and enables greater accuracy (see Figure 4.29). Nevertheless, finger vein systems do have at least one specific, incontestable advantage over hand vein systems—numerical superiority. Fingers have a 5:1 numerical advantage, and 10 fingers offer a variety of application options. Most system providers suggest that each individual register a minimum of two fingers on each hand during the enrollment process. This provides flexibility in case either hand sustains an injury. It also enables an authorized user to overcome false

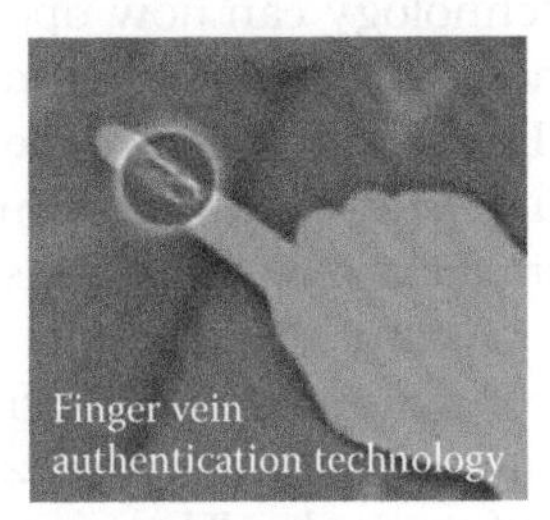

FIGURE 4.28 Finger vein technology. (Courtesy of Hitachi Ltd.)

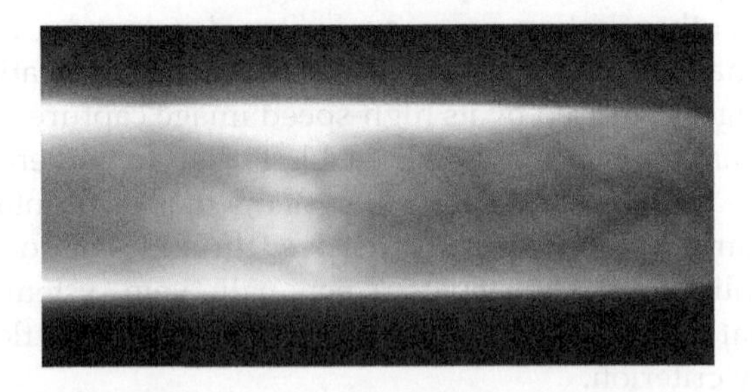

FIGURE 4.29 Near-infrared image of finger veins. (Courtesy of Bionics Co., Ltd.)

rejects by simply using another finger; if registered properly, virtually all individuals are authenticated on additional attempts.

Hitachi Ltd.

Since 1997, Hitachi has been developing finger vein authentication technology, which uses the finger vein pattern obtained from passing light through a finger as a key. Its first commercial systems were deployed in 2004. In 2005, a grip-type finger vein authentication technology was developed, enabling a door to be opened simply by gripping the handle. Since then, Hitachi has been working to develop an even more compact system to extend market applications.

As illustrated in Figure 4.30, near-infrared light generated by a bank of LEDs penetrates the body tissue and is reflected in the hemoglobin in the blood. A CCD camera (which uses a small, rectangular piece of silicon to receive incoming light) captures the image of the vein pattern through this reflected light. Image processing constructs a finger vein pattern from the camera image. This pattern is compressed and digitized so that it can be registered as a template for biometric authentication. Within a split second, the finger vein system filters the digitized image, produces a template or digitized image that it compares to the stored template of the user, and determines whether there is a match, using pattern-matching techniques.

Hitachi has delivered to the market a variety of finger vein scanning devices, depending on the application. Hitachi's initial successes with VPR technology were with Japanese financial institutions and their ATM systems. Hitachi-Omron Terminal Solutions (HOTS) designed a physical

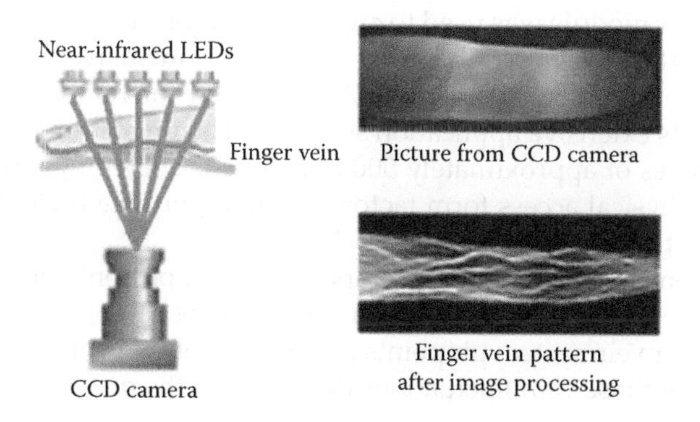

FIGURE 4.30 Technology of finger vein pattern recognition. (Courtesy of Hitachi Ltd.)

FIGURE 4.31 Finger vein module. (Courtesy of Hitachi-Omron Terminal Solutions.)

access module to be embedded in ATMS, kiosks, and turnstiles. Shown in Figure 4.31, this module is highly accurate and has a small portable form factor that enables it to be mounted inside a variety of larger devices. It has been successfully used at 75% of the bank branches in Japan, including such institutions as Japan Post, Mizuho Bank, Sumitomo Mitsui, and Resona Bank.*

A similar module was used to create stand-alone physical access readers such as the one shown in Figure 4.32. The VeinGuard physical access reader has exceptional security, is Web-enabled for remote diagnostics, and supports extensive application software. The finger vein readers support templates of approximately 500 bytes each. They range in size from the larger physical access form factors to small portable finger vein readers, such as the one shown in Figure 4.33.

Variations in environmental temperature or a person's blood pressure can sometimes cause fluctuations in the width of the blood vessels. The Hitachi finger vein system pinpoints the position of the center of each blood vessel so that those vein fluctuations do not affect the matching procedure.

* This is according to a survey of all Japanese banks taken in September 2006.

FIGURE 4.32 VeinGuard physical access finger vein terminal. (Courtesy of Hitachi America Ltd.)

Hitachi has developed a grip-like finger vein verification technology that can authenticate a vein pattern on the back of a finger, which verifies the identity of the person attempting to enter a room, home, or vehicle. Authentication occurs as part of the action of opening a door. This reaches a new level of convenience and obviates the need to carry keys or perform some special task prior to entry. An authorized person has only to grip the handle to have his identity confirmed. Indeed, this grip-like application can effectively incorporate personal authentication in the natural motion of opening a door (for example, home, office, or automobile doors).

This differs from Hitachi's basic finger vein products, which image the vein pattern on the palm side of the finger. When someone performs a gripping action, such as that shown in Figure 4.34, the finger veins on the palm side have a tendency to compress and distort the standard vein pattern, making it difficult to obtain a consistent, stable vein image. On the other hand, the stretching of the veins on the back side of the finger enables capturing a consistent image by locating light emitters on the handle component in proximity to the back of the fingers when gripping. It is also helpful to have a door handle system designed to naturally guide the fingers into general position, further increasing consistency. Finally, there are compensation techniques that can further increase accuracy by

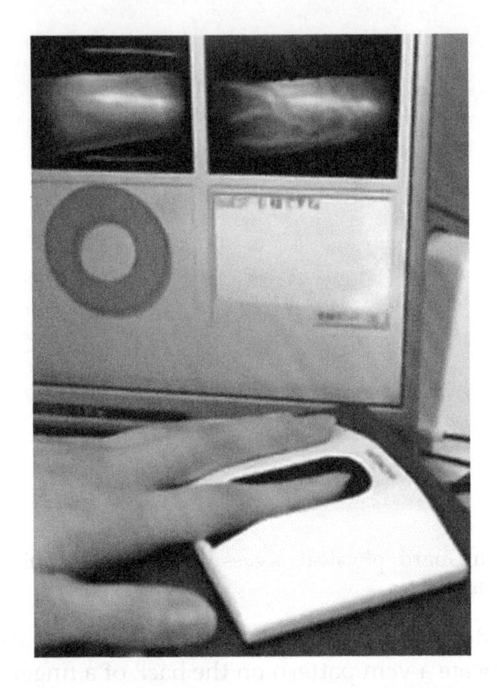

FIGURE 4.33 Portable finger vein system. (Courtesy of Hitachi Ltd.)

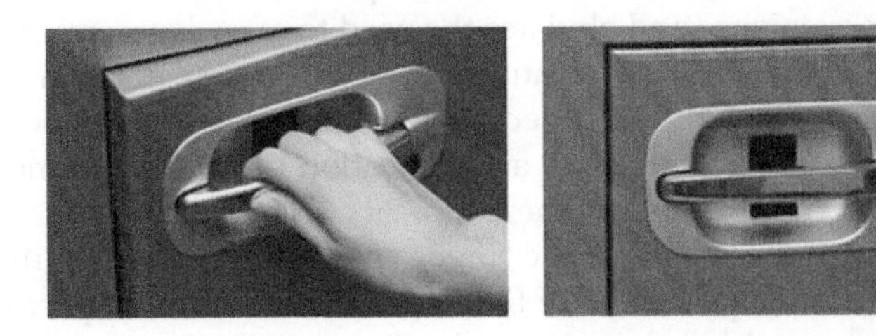

FIGURE 4.34 Grip-type finger vein authentication technology. (Courtesy of Hitachi Ltd.)

stabilizing the authentication results. Grip-like technology enables personal authentication through the natural movement in opening a door for a car, business, or home.

Bionics Company, Ltd.

Bionics Co. Ltd., based in Osaka, Japan, was established on January 29, 2001. Bionics has developed a highly reliable and secure vein authentication

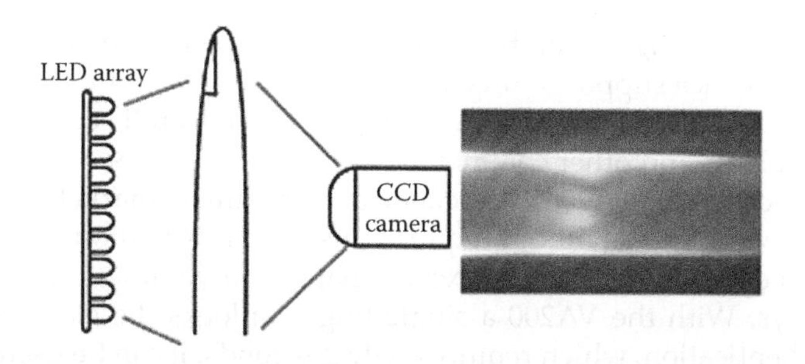

FIGURE 4.35 Near-infrared image of finger. (Courtesy of Bionics Co., Ltd.)

system focused on the palm-side of the finger. In 2006 Bionics formed Bionics America Inc. to address opportunities in the United States.

Bionic's Blood Vessel Authentication System (BVAS) recognizes a person by analyzing the uniquely random pattern of the finger veins. The automated method of finger vein recognition was developed to address inherent disadvantages in other biometric modalities. As Figure 4.35 illustrates, BVAS uses permeating, near-infrared light and a high-resolution CCD camera to capture the transilluminated image of the blood vessel pattern in the finger. An algorithm extracts characteristic points from the digitized image, and the system achieves individual authentication by comparing the sample pattern results against the stored enrollment template. Bionics has published its accuracy levels as a false acceptance rate of less than .0001% and a false rejection rate of less than .01%.

Bionics has two basic products: the VA100 for commercial and industrial sites, and the VA200 for condos and apartments.

Originally designed for time and attendance applications, the VA100 is a two-factor device that uses a PIN and the unique image of the vein pattern in a finger for enrollment and authentication. Figure 4.36 is a photo of the VA100 scanner. Systems that use it normally enroll up to 800 users

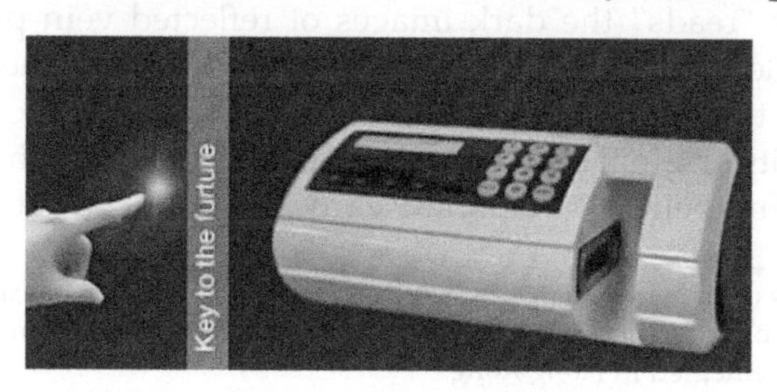

FIGURE 4.36 VA100 finger vein scanner. (Courtesy of Bionics Co., Ltd.)

in a standard configuration but can optionally accommodate up to 3,000 users. The device supports three access levels: users, administrators, and supervisors. The VA100 is deployed as part of an overall security system at power plants and other commercial and industrial sites.

The focus of the VA200 is the condo and apartment market. The VA200 enables keyless entry without requiring use of a PIN (although it is available as an option), and it offers exceptional security without the worry of losing keys. With the VA200 a single finger unlocks the electronic lock after authentication, which requires only 2 seconds. It can be used in conjunction with room access control software, which can display the status of a room in real time. The VA200 can be used with smart cards or remote access via the Internet. Additionally, it can be deployed as a stand-alone system or integrated into an existing or developing system.

BVAS has performed well in Japan. Bionics finger vein readers are deployed in Japan at the Nuclear Cycle Development Institute and the Technical Research Institute of Clinical Lab. In North America, Bionics has deployments at Luis Munoz Marin International Airport (Puerto Rico); U.S. federal banks; and at various high-security private and commercial establishments and residences. Figure 4.37 lists some completed projects in Japan at both public and commercial facilities.

Sony Corporation

In February 2009, Sony introduced its new finger vein authentication solution for mobile devices. It calls this new product Mofiria. This unusual name appears to be a contracted form of Mobile, Finger, and Integration Authentication. The CMOS (complementary metal-oxide semiconductor) sensor in Mofiria diagonally captures the vein patterns inside one's finger. It uses a "reflecting scattering light method" with a special algorithm that converts the vein pattern to a plane layout. By directing near-infrared light emitted from a bank of LEDs at the finger vein patterns at an angle, Sony's CMOS sensor "reads" the dark images of reflected vein patterns. Sony uses a multiple filtering process to extract the image of the finger's vein pattern. Prior to template storage, the extracted pattern is compressed to about 10% of its original size, reducing the template to such a level that it can be stored in a FeliCa* card or SIM† card. The template data is encrypted

* FeliCa is Sony's contactless smart card used in electronic money applications. The name is an abbreviation for Felicity Card. It was first used in the Octopus smart card system, which was implemented in Hong Kong.

† SIM is an acronym for subscriber identity module. It is a small electronic card inserted in mobile phones, providing a unique ID to the phone.

Completed Projects

Public and Commercial Facilities in Japan

Approximately 1,000 Installation Sites

Bionics

FIGURE 4.37 Completed projects, public and commercial facilities in Japan. (Courtesy of Bionics Co., Ltd.)

prior to storage, and Sony enables the user to select DES (Data Encryption Standard), AES (Advanced Encryption Standard), or Sony's "Clefia" as encode methods. The system enables on-device template matching.

By placing the LEDs and CMOS senor on the same side, Sony was able to achieve a smaller scanner system. Mofiria is very compact as its form factor was designed for mobile devices such as PDAs, laptops, and mobile telephones. Sony appears to have built a fairly simple and straightforward device structure with an innovative but uncomplicated method to capture vein pattern data. It offers a user-friendly interface that is purported to automatically account for the finger's position on the scanner.

Sony has retained much of the accuracy associated with existing finger and palm vein scanner devices. It uses an Intel 2.8 GHz microprocessor for notebook PCs and an ARM9 150 MHz core-based microprocessor for mobile phones; and it achieves a .015 seconds and .25 seconds throughput speed, respectively. Sony's purported false reject rate is 0.1% and false acceptance rate is 0.0001%. Although both accuracy measurements are nearly an order of magnitude less accurate than the Hitachi finger vein systems, Sony is serving a different market. It is unlikely that a laptop, PDA, or mobile telephone would be used by more than a handful of authorized individuals. And this accuracy level is much greater than that achieved by most fingerprint systems that are embedded in laptops today.

Clearly, Sony has targeted Mofiria at mobile devices. It has enabled each finger that is registered to the mobile device to associate with a different application. So one's index finger can log onto a PC, one's middle finger can initiate the browser, and one's ring finger can launch presentation software—all in accordance to the wishes and imaginations of the users. Mofiria can also require multiple finger verification to increase security. What Mofiria clearly hopes to facilitate is the storage of important or sensitive information on a mobile device with confidence in the security of the data. Moreover, Sony is expected to target gateway security applications with strong authentication for Internet and corporate intranet applications.

Sony has not announced specifics regarding the launch of its Mofiria product, but it has indicated that the product will be released for sale within its 2009 fiscal year (which ends March 31, 2010). Sony did not provide any information regarding its pricing or its distribution channels.

5

Vein Pattern Recognition Applications

Biometrics has been employed in a variety of applications. Most people associate biometrics with security applications such as physical and logical access. Indeed, physical access to buildings, laboratories, data centers, and other secure facilities is a highly visible, traditional application for biometric technology. Moreover, logical access to commercial and corporate networks, personal computers (PCs), financial accounts (for example, automated teller machines [ATMs]), and other virtual systems is a strong and rapidly growing application. Clearly, vein pattern recognition (VPR) systems can readily address physical and logical access control applications. However, there are a growing number of other applications such as workforce management—a significant headache for many businesses, and one that biometrics can solve with efficiency and convenience. Another key application is membership, whereby a member of a given organization needs to verify his identity to access the benefits of membership. Biometrics can determine if an individual is already part of a database, such as someone seeking a social service benefit, driver's license, or national ID. Finally, biometrics is extremely useful in accountability applications whereby one needs to authenticate an identity, such things as boarding a commercial aircraft, ship, bus, or train; maintaining a chain of evidence; or signing for a classified document. Although these broad categories are by no means the only appropriate applications for biometric technologies, they are the primary ones in use today.

Early commercial success stories for biometrics were for applications with provable returns such as physical access control, password reset (e.g., logical access), and time and attendance. It is also important to understand that VPR systems, like other biometric products, are most efficiently used not as security add-ons or even stand-alone systems, but as key components of a comprehensive security solution. That is why skillful system integrators play such a key role in VPR solutions. Implementing unique security that is difficult to use may not be acceptable to the user population who will simply not use it or find ways to work around it.

This chapter describes the six major applications groups for which VPR-based biometric applications are used: physical access, logical access, workforce management, memberships, accountability, and embedded systems. Vein pattern recognition biometrics is an important component to a total solution for combating fraud, identity theft, and unauthorized access. The technology is very reliable and it reveals limited performance degradation in harsh environments such as construction sites or heavily trafficked venues such as military posts or schools.

PHYSICAL ACCESS

Physical access control refers to the policies, procedures, and technical controls that regulate access to a building, a room, an ATM,* or other physically defined structures. Its protection is called physical security. In many minds, the term *physical access* conjures images of cards and keys. Used alone, badges and proximity cards are obsolete, and in the next decade one's own body will replace physical keys. In a growing number of situations, biometrics is being used for physical access control applications at entrance and exit points of a building, floor, or office to restrict access to authorized individuals. Figure 5.1 shows a vein pattern reader controlling access to a data center. Increasingly, enterprises are choosing to incorporate biometrics as part of their overall security programs in a layering process. Layering refers to the use of both biometrics and other nonbiometric security mechanisms such as physical tokens (RFID [radio frequency identification] or smart card–based) and personal identification

* ATMs are considered extensions of the bank's brick and mortar. They provide physical access to cash, and they generally use the same form factor as other physical access devices; for those reasons, I have placed ATM access in physical rather than logical access control.

FIGURE 5.1 VPR reader used for data center access. (Courtesy of Identica Corp.)

numbers (PINs) or passwords. Layering supports heightened security and enables greater flexibility in the use of authentication technologies.

Physical access is the most commercialized of the biometric applications, and it claims the largest market share among horizontal applications. Hand geometry and fingerprint scanning are the most commonly deployed physical access solutions. However, the market has been rapidly shifting to other biometric modalities such as iris scanning and VPR because of requirements for greater accuracy, usability, and user acceptability. These criteria are key people considerations. Before an enterprise requests its employees to provide their biometric identifiers to access the enterprise resources, it might be prudent to consider the perceptions of its staff and other stakeholders. Enterprises can reduce user anxiety by implementing biometric solutions that address people issues including privacy, the need for hygienic solutions, and other cultural concerns.

The current most active markets for biometric-based physical access control systems are government, financial institutions, and high-security venues (for example, power utility sites, data centers, and so forth). Physical access control applications are not limited to high security installations or restricted access venues, but are thriving at apartments, villas, and luxury hotels. VPR technology can be integrated into physical access control of electronic door lock systems. Most VPR access control devices consist of a vein pattern image sensor, a small display, and an optional keypad, as well as a control unit that executes the authentication processing and forwards the unlock instruction. VPR systems can connect directly to the electronic lock, or to a security system that controls the lock.

Ports

Airports, seaports, and land border-crossing points can differentiate themselves in a very customer-friendly manner by providing increased security and improved convenience through biometrics. Airports and seaports are introducing this technology at customs, immigration, and quarantine areas. Turnstile access is easily retrofitted with VPR scanners (see Figure 5.2). Border crossings tend to use a multiple-biometric approach for large-scale applications. One example where VPR technology is screening passengers or staff is South Korea's Inchon International Airport; another is Canada's Port of Halifax.

The Port of Halifax is one of the largest seaports in North America. It covers 20 acres of land and it serves as a gateway to eastern Canada and the northeastern United States. More than 4,000 dockworkers pass through approximately 2,200 access points located throughout the port, using Identica's back of the hand VPR technology. In making its decision regarding which biometric method would be best, the port authority considered the severe cold and other weather factors as well as worker acceptance, especially their privacy concerns. The port rejected iris scanning and fingerprint scanning due to perceived intrusiveness and privacy implications of those biometric modalities, respectively. The port considers VPR technology to be nonintrusive and privacy enhancing. Additionally, the port issued each of its workers a smart card for storage of his or her VPR template. To shield the biometric scanners from the harsh weather, the port outfitted them with a weather shield, an automated, heated outdoor enclosure, as shown in Figure 5.3.

FIGURE 5.2 VPR reader used at outdoor turnstiles. (Courtesy of Identica Corp.)

FIGURE 5.3 Enclosed VPR reader, weather protected. (Courtesy of Identica Corp.)

Financial Institutions

Most financial institutions are security conscious and are actively soliciting and implementing multifactor authentication solutions. Swiss and American banks have introduced access control to vaults and other key areas with Identica's back of the hand scanners replacing keys and combination code entry. Banco Bradesco, Brazil's largest private bank, has incorporated Fujitsu's palm vein technology into its ATMs, as have a significant number of other banks throughout the world.* Financial services, such as ATMs, payment terminals, cashless systems, automated check cashing, and so forth, are applications that are being evaluated for biometric systems outfitting. Banking is broadening support for home-based solutions, which will increase the need for unique biometric identifiers for logical access to the user's banking information. Like many enterprises, financial institutions are actively seeking opportunities to provide new or improved services to their customers. Biometrics helps to achieve that goal.

SunFirst is the fastest-growing community bank based in St. George, Utah. It has a reputation as a technologically advanced financial

* Some notable banks in Japan that employ Fujitsu's palm vein system in ATMs include: Suruga Bank, Bank of Tokyo-Mitsubishi, Hiroshima Bank, Bank of Ikeda, and Nanto Bank.

institution. It chose Identica's hand vein system to protect entrance into the bank's IT (information technology) data center. The use of this VPR system provides an extra layer of protection for the bank and its customers while keeping the data center's management constantly aware of who enters and exits during any particular time period. SunFirst chose its VPR solution because it was accurate, easy to use, and had a quick and simple enrollment process.

What occurred in Japan with the mass deployment of vein pattern technology for its retail banking sector was truly amazing. In the early 2000s, Japan experienced increased incidents of illegal ATM withdrawals and other forms of financial fraud from stolen cards and the practice of skimming.* By 2005 the funds stolen exceeded US$8 million, and these fraud cases generated sensational stories in the media. Unlike in the United States, ATM cards in Japan had no withdrawal limit prior to 2005,† and Japanese banks did not guarantee loss due to theft.

The Japanese government came under strong pressure to combat the growing fraud problem. In response to this growing concern, the Japanese Diet passed a new law in August 2005, referred to as the Depositor Protection Law,‡ which took effect on February 10, 2006, and provided some protection for depositors. This law indemnified depositors for damages incurred (e.g., loss of funds) by identity thefts. However, it applied only to ATM withdrawals that used forgery and theft of bank cards. In this situation, the liability for the loss transferred to the banks, but only if the depositor exhibited no negligence such as placing a recorded PIN with his ATM card. There remained among the Japanese people considerable concern regarding the security of the banking system, a concern shared by the Japanese banks, which now had their own liabilities to consider. Therefore, to meet the demand for more secure ATM banking, the nation's banks and depositors rapidly embraced biometric solutions. Indeed, the process had begun even before the new law passed.

Because of a cultural bias against fingerprint technology and a preference for hygienic solutions, Japanese banks began to rapidly adopt VPR. Japan's Suruga Bank first used the palm vein pattern recognition in July

* Skimming is electronically copying data from one card to the next.

† In 2005 Japanese banks began to allow depositors to set their own withdrawal limits on their accounts.

‡ The legislation was actually called the Law Concerning the Protection of Depositors from Illicit Deposit Withdrawals Using Counterfeit/Stolen Cash Cards through ATMs, but it is commonly referred to as the Depositor Protection Law.

FIGURE 5.4 VPR reader used at an ATM. (Courtesy of Hitachi-Omron Terminal Solutions.)

2004. Japan's Bank of Tokyo-Mitsubishi quickly rolled out Fujitsu's VPR technology to its bank customers on 5,000 ATMs beginning in October 2004. The bank provided smart cards issued to its customers for credit and debit applications, and for template storage, and called its product the Super IC Card Tokyo-Mitsubishi VISA. Hitachi introduced its finger vein technology with smart cards the following March and soon had implementations at a substantial number of institutions including the Japan Post,* Mizuho Bank, and Sumitomo Mitsui Bank (see Figures 5.4 and 5.5). With the passing of the Depositor Protection Law, VPR implementations at ATMs accelerated. Today, over 92% of all bank branches in Japan use VPR technology.† As a result of the successful adoption of the vein pattern technology by the Japanese banks, there was a significant reduction in financial fraud.

For the depositors, the biometric solution was voluntary, but the vast majority of people quickly embraced the solution. Each depositor was given a smart card and then enrolled in the vein pattern system the bank had selected. Each depositor's vein pattern enrollment template was stored on his or her smart card, which is protected by the user-selected PIN. In

* Japan Post supports the largest network of ATMs throughout Japan.

† As of the summer of 2007, there were 48 financial institutions representing 26,548 branch locations using the finger vein technology, giving Hitachi a significantly large market share.

FIGURE 5.5 Another view of VPR reader in an ATM (Courtesy of Hitachi-Omron Terminal Solutions.)

Japan, the banks employ three-factor security: a PIN, a smart card, and a biometric identifier.

As illustrated in Figure 5.6, when an individual uses an ATM, he places his smart card in the ATM and then enters his PIN. The PIN authenticates the individual to his smart card. He then places his hand or finger on the ATM's biometric reader. The VPR device embedded in the ATM scans the depositor's palm or finger, dependent on the system used, and creates a "live" vein pattern sample template that is transferred to the depositor's smart card. Next, the new sample template is then matched with the enrollment template on the smart card itself. Once a match occurs the smart card verifies the depositor's identity to the ATM network. The on-card matching and decision making reinforces the security of the depositor's vein pattern and protects his account.

Buildings

In Singapore, a number of well-known buildings such as IBM Singapore, Caltex Tower, and Mizuho Bank have adopted Hitachi finger vein systems

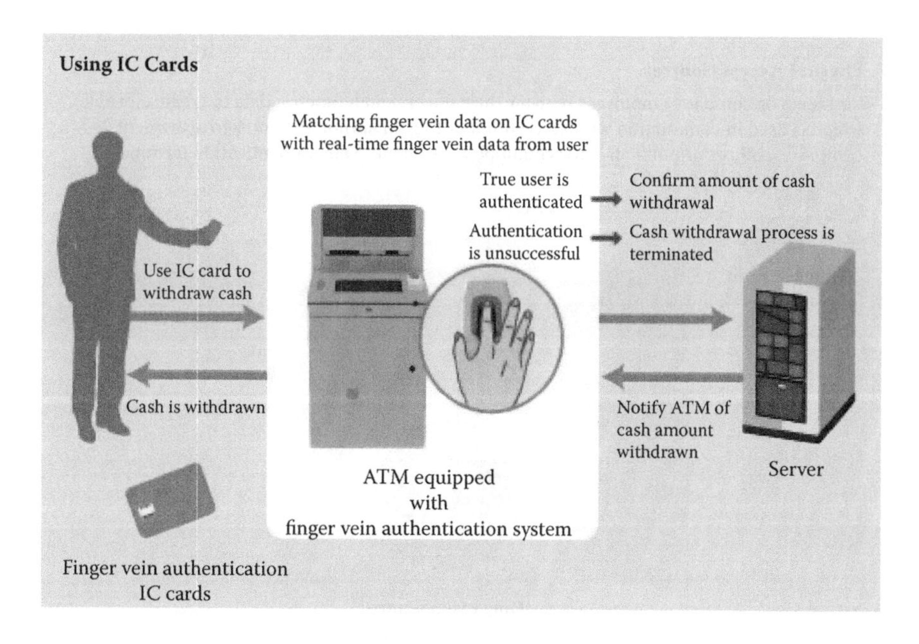

FIGURE 5.6 Schematic using smart cards with a VPR reader at an ATM. (Courtesy of Hitachi-Omron Terminal Solutions.)

as key components to their respective physical access security systems. Figure 5.7 illustrates building access control.

The University of Tokyo Hospital has added Fujitsu's palm vein scanners to secure its room access, replacing its fingerprint-based system. A contactless palm vein scanner was installed at the entrance of each room to prevent unauthorized entry.

The majority of hotels, motels, and apartments have moved away from physical keys to electronic ones, usually magnetic stripe card based. Access to buildings is now migrating to proximity cards and biometrically enabled smart cards, or to biometrics alone.

High-security venues might require additional support for protected resources, and thus may continuously monitor individuals with multiple verifications. For example, VPR can help authorized courtroom staff to have convenient access to facilities, rooms, or areas in a secured building. Forgotten access codes or disclosure of PINS (inadvertent or deliberate) become nonissues when VPR requires the individual to be verified to be present. This protection can extend to court information systems.

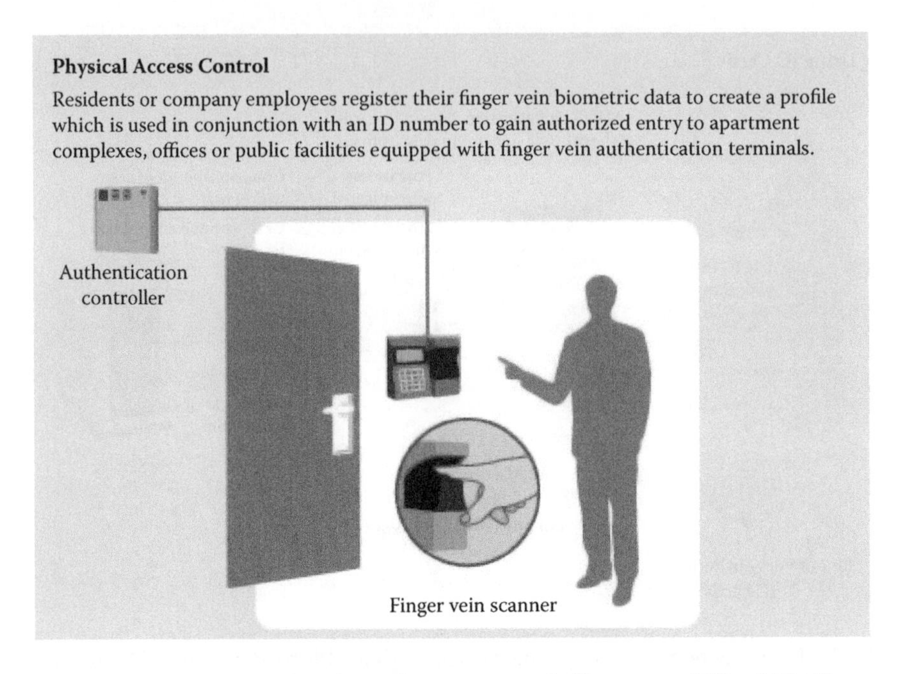

Physical Access Control
Residents or company employees register their finger vein biometric data to create a profile which is used in conjunction with an ID number to gain authorized entry to apartment complexes, offices or public facilities equipped with finger vein authentication terminals.
Authentication controller
Finger vein scanner

FIGURE 5.7 Schematic for physical access control. (Courtesy of Hitachi Ltd.)

Safe Deposit Boxes

A safe deposit box serves as a personal private vault for storing important papers, jewelry, and other valuables. Safe deposit boxes are all about privacy and convenience while protecting one's valuables in a secure, loss-resistant environment. In a major departure from the traditional card token and signature access, biometric technology is changing the way in which safe deposit customers gain access to their boxes. At banks, the elimination of the manual steps in retrieving the signature card to provide customers access to their property saves valuable time, and customers appreciate the speed of access. Biometrics cannot only authenticate the visitor but can also retain an electronic log of everyone who enters or exits the safe deposit box vault.

Many financial institutions and numerous independent safety deposit vault operators are now using biometric vault-entry systems to enable convenient and authorized access to safe deposit boxes. Customers are also very aware of the dangers of identity theft. The use of biometrics simply

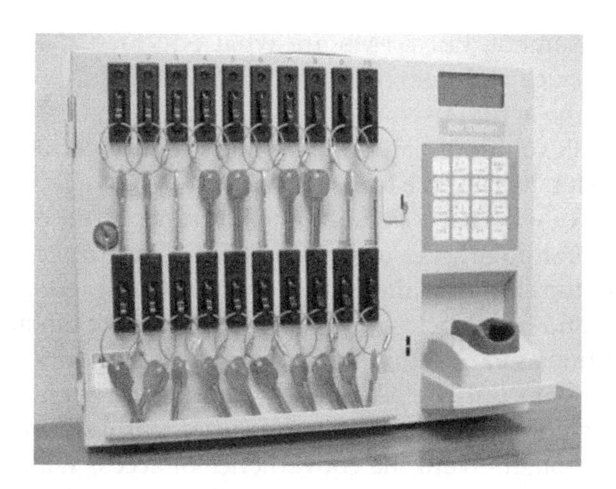

FIGURE 5.8 Key management system with a finger vein reader. The outer case metal enclosure was removed for demonstration purpose. (Courtesy of Hitachi-Omron Terminal Solutions.)

provides another layer of security to protect the valuables and identities of customers.

Safe deposit boxes are not limited to banks or specialty vaults at independent safe deposit enterprises. Hotels usually accept and deposit valuables. In the United States the innkeeper liability laws regarding loss of guest valuables may not always apply to in-room safes. Moreover, a biometric access would be more convenient than remembering a code or carrying a key. Figure 5.8 depicts a key management system protected by finger vein pattern technology. With this system an authorized individual inputs his PIN via the device's keypad, and then presents his finger to the finger vein reader. The key management system enables authorized people to access specific keys to restricted areas while maintaining access records, which support the concept of nonrepudiation.

The use of biometric identifiers complements the security of the safe deposit vault, and they increase convenience and privacy. Vein pattern recognition technology is the strongest choice for supporting this application.

Self-Service Solutions

The concept of self-service is well established, but achieving its promise can be illusive. Self-service is evolving. Vending machines led the self-service movement, but ATMs are what educated the public on the value of self-service technology. Most ATMs can identify a card and its owner, but not the actual user. Self-service starts with customer authentification, and VPR solutions can improve that identity verification process, resulting in faster, hassle-free transactions.

Regardless of whether a person owns or rents his home, chances are he has run out of storage space at some point and has turned to self-storage as a solution to the cluttered alternative. One's self-storage space is usually self-contained and fully enclosed with 24-hour access. Self-storage is often a secured solution as each individual provides his own lock, meaning that the storage unit can only be accessed by the person with the key to the lock. Biometric solutions are gaining ground with self-service storage because the customers want the convenience of access without having to remember their keys or having to continually transfer their keys to other individuals that they have authorized. College students have particularly high incidents of replacing their locks (after bolt cutting them off) each new year, as they lose their keys or forget the lock combinations.

Self-service storage is not limited to an empty cubicle-like space where people can simply store their treasured possessions. For wine enthusiasts who need complete wine storage systems, there is wine cellar storage at a stable temperature and humidity. Cigar aficionados might want their special supply of cigars to be safely stored in local humidors so that they can protect their perishable investments. Some people store their audiovisual records for long-term preservation in special audiovisual vaults that deter deterioration.

Airports have turned to self-service to resolve passenger congestion and to meet security demands. To lessen delays at the various airport touch points—check-in, security, baggage collection, and airline boarding—airports and airlines have embraced self-service kiosks, remote passenger check-in, luggage drop-offs, and biometrics. For example, Air France may be introducing the next level of self-service by enabling passengers to use biometrics to confirm their identities and to then print boarding passes for use at the boarding gate.

Car rental companies have enabled self check-in at car rental kiosks. Car rentals are a market segment where speed and identity authentication are crucial. Some companies are still focused on the magnetic stripes on driver's licenses, while others have chosen biometric identifiers, which increase productivity and security, and speed up the entire process.

Another obvious use of biometric-based self-service is grocery shopping. The modern supermarket already operates in a self-service format.

Shoppers select their items as they tour the store and fill their shopping baskets. The check-out process is just another facet of what is already a self-service model. As people become more comfortable with the check-out technology and as retailers continue to face labor challenges and strive for greater efficiencies, automated solutions such as customer self-service just make sense. However, biometrics could truly enhance the shopping experience and personalize the self check-out by enabling faster service and support applications such as self-returns and supervisor overrides, significantly reducing the need for store clerk interventions.

There is an immense opportunity to drive transactional self-service processes throughout enterprises. It is all about empowering staff and customers to serve themselves. By automating administrative functions and empowering employees toward secure self-direction and independent action, an enterprise can acquire a strategic advantage vis-à-vis its competition. Organizations such as ii2P (Integrity, Innovation, People—based in Southlake, Texas) promote self-service initiatives for identity verification applications,* including smart card-based and biometric-based solutions. A successful self-service approach can slash costs from a service interaction, whether the purpose of the interaction is to fix a problem or update personal data. And the prerequisite for a successful self-service solution is authenticating the identity of the individual serving himself.

LOGICAL ACCESS

Logical access control refers to the policies, procedures, and technical controls used in information systems. The protection of logical access is called information security. Logical access applications include system and application log-on, workstation and network access, single sign-on, data protection, remote access, and Web applications. Today, the key markets for biometrics-based logical access are government, electronic banking, health care, and enterprise-wide networking.

Computer and security solutions encompass PC or service log-on, e-commerce, and workflow applications such as printer and copier control. The incorporation of a biometric solution can help protect the sensitive

* Identity and access management (IAM) merges business processes, security policies, and a variety of identity verification technologies (e.g., biometrics, smart cards, public key infrastructure [PKI], one-time password [OTP], etc.) to help enterprises manage digital identities.

information available through computer networks as well as protect laptops from unauthorized access. Companies like Aeon Credit Services in Japan and several Japanese government agencies use USB-connected finger vein readers from Hitachi for PC logical access.

Health Care Services

Health care facilities are increasingly using biometric solutions to verify patient identity and to ensure correct treatment, medication, and proper use of personal data. Carolinas Healthcare System (CHS), based in Charlotte, North Carolina, uses Fujitsu's palm vein technology, PalmSecure™, to more efficiently streamline patient registration and check-in, as well as to safeguard patients' electronic medical records (EMRs). CHS is the largest health care system in the Carolinas and is one of the largest nationally. CHS associates each patient's registered palm vein template with the patient's EMR. When the patient checks in to the hospital, he simply scans his palm vein pattern, which serves as an individual identifier and is linked to his EMR. Next, the patient's demographic and health care information is automatically populated on the registration screen. This technology helps protect patients from identity theft and insurance fraud, and is in compliance with HIPAA.* It significantly streamlines the check-in process and enhances the patient's experience while protecting the patient's privacy. Additionally, CHS developed a unique hand guide for the image sensor, which incorporates a "pediatric plate" to adopt the guide for young children, enabling CHS to accommodate its entire patient population.

BayCare Health System, a large full-service health care provider based in Tampa, Florida, uses the Fujitsu PalmSecure palm vein scanners at nine BayCare hospitals and multiple outpatient clinics and laboratories for patient registration and identity verification processes. By integrating Fujitsu's biometric solution into its IT systems, BayCare has eliminated its duplicate medical record creations because only one EMR can be linked to a VPR template. BayCare can securely and accurately verify a patient's identity even if he or she is confused or is unconscious, while safeguarding the patient's privacy.

---

* HIPAA is an acronym for the Health Insurance Portability and Accountability Act of 1996 (August 21), Public Law 104-191. HIPAA published rules that (1) provided standards for electronic patient health, administrative, and financial data; (2) provided for unique health identifiers for individuals, employers, health plans, and health care providers; and (3) provided security standards protecting the confidentiality and integrity of "individually identifiable health information," past, present, or future.

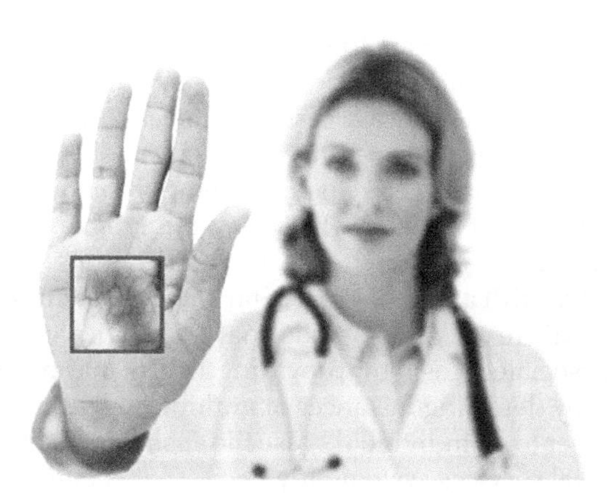

FIGURE 5.9 Fujitsu's palm vein system is a health care success. (Courtesy of Fujitsu Ltd.)

Figure 5.9 is part of a Fujitsu advertisement representing its successful marketing in the health care arena. For the health care sector, VPR offers key advantages: improved security, efficiency, safety, convenience, and privacy protection used to register patients (check-in), maintain access to specific laboratories or exam rooms, and secure participation distribution.

Electronic Benefits Transfer

Currently, governments are not generally using biometric solutions in determining eligibility for or delivery of government benefits. In the United States biometrics is notably absent in many programs such as most WIC,* TANF,† Medicaid/Medicare,‡ and SNAP§ implementations, even though it is well known that these programs are beset with fraud. Most federal government agencies and state governments have been content to

* The Special Supplemental Nutrition Program for Women, Infants & Children (known as WIC) and SNAP are nutritional safety nets for families living in poverty in the United States.

† TANF is an acronym for Temporary Aid to Needy Families, a cash assistance program for low-income families (also known by the colloquial term *welfare*).

‡ Medicare and Medicaid were created by Titles 18 and 19, respectively, of the Social Security Act passed by Congress in 1965. Medicare supports the medical needs of the elderly or severely disabled, whereas Medicaid provides public assistance medical care for low-income people and families.

§ SNAP is an acronym for Supplemental Nutrition Assistance Program. It was formerly referred to as the Food Stamp Program.

sit on the sidelines and evaluate the success of other programs regarding the potential benefits that biometric technology can provide, including reduced fraud, improved service access, and lower costs. These programs are federally funded but state operated.

While most states have been reluctant to employ biometric verification to their electronic benefits transfer (EBT) programs, a handful of states do use them,* including Texas, which uses biometrics in its Texas Medicaid Integrity Pilot (TX MIP) and in its Lone Star Image System (LSIS) for food stamps and TANF. Texas understands that biometrics can help control fraud and its associated costs by providing irrefutable proof of identity verification and by detecting instances of multiple registrations. Linking a biometric identifier to an individual's case file ensures that each person can enroll in the social services system only once. That is one of many factors that helped compel Texas to implement a biometric-based identity management solution. Similarly, in Los Angeles County, California, the use of biometrics to verify identity is a key part of the LA AFFIRM program.

Not only does a biometric solution help prevent duplicate participation (e.g., same person but different names and addresses, also called double-dipping) in a given program, but it reduces the rising administrative costs that accompany such fraud. Moreover, government program directors usually strive hard to preserve program integrity, so the removal of perceptions of fraud and abuse in their programs generally would be a welcome development. The use of biometric-based identity management programs can ensure the programs' stakeholders that issues of fraud and abuse are being appropriately addressed and thwarted. Additionally, government agencies can realize additional nonfinancial benefits through the reduced duplication of effort, streamlined processes, and the ability to better measure the program benefits for the recipients. Using biometric authentication at the point of service delivery could significantly improve data integrity. Positive identity verification can enhance service delivery by improving access and entitlement. Resulting process improvements can more efficiently make use of public funding by reducing staff time, eliminating redundancies, and reducing rejected claims based on eligibility.

Even though many organizations continue to separately administer physical security and information security, they are remarkably similar, even to the task level. In the United States, the Homeland Security

* States using biometric identifiers in large-scale social services programs include Arizona, California, Connecticut, Massachusetts, New York, and Texas.

Presidential Directive 12 (HSPD-12) established the requirement for a common identity credential to garner physical or logical access to government resources.

Strong Authentication

To increase business agility, competitive enterprises require innovative methods to manage secure access to information and to applications across multiple systems. As discussed in Chapter 1, passwords and single sign-on solutions are not enough; strong authentication is required. Companies are turning to a broad variety of strong authentication options to stimulate user adoption into existing workflows. Strong authentication is driving improved cost efficiencies, heightened security, and higher employee productivity across variegated organizations.

Fujitsu created PalmSecure LOGONDIRECTOR to address the needs of enterprises requiring biometric support for strong authentication. Indeed, Fujitsu integrated its VPR capabilities with PassLogix's v-GO Access Accelerator* suite for enterprise single sign-on (ESSO). Fujitsu's VPR solution seamlessly integrates with third-party SSO solutions. Using the Fujitsu palm vein solution, users no longer have to remember passwords at all. In other IT environments, multiple users may need to access a single PC; in such a scenario, the use of a VPR identifier can enable tracking of individual activity and can limit user access to only the applications and files for which the individual is authorized. VPR technology can enable convenient and secure identity verifications to provide strong authentication in PC log-in as well as reduce the complexity of managing multiple passwords.

WORKFORCE MANAGEMENT

Workforce management refers to employee safety, scheduling, and productivity and includes time and attendance (T&A) applications such as scheduling, attendance, payroll, and individual job costing. For years, T&A was dominated by time cards, time sheets, or time clocks with a lot of headaches and high costs associated with supporting traditional password and token-based systems.

* The v-GO suite is an integrated set of single sign-on authentication and provisioning enablement solutions.

One focus of existing biometric T&A products is to eliminate the proxy transaction referred to as "buddy punching," a phenomenon whereby well-meaning "buddies" clock in a friend who is running late or not coming in at all, compromising the system. A biometric T&A system eliminates such proxies.

Biometric-based T&A systems fix the issues associated with the manual time clocks and time cards, as well as support additional applications. Most can automatically classify the hours collected into such categories as regular time, overtime, and double time. They can calculate gross payroll and interface to back office systems and can manage all scheduling, including administering attendance policies. Perhaps their true value is ensuring compliance to local, state or province, and national laws by tracking and applying complex work rules. T&A systems measure workforce attendance and productivity, enabling analytic review of the enterprise performance.

Some of the older workforce management systems, which use hand geometry and fingerprint systems, have relatively high false reject rates. Vein pattern recognition technology is generally more accurate, more hygienic, more secure, and more convenient for users. Most VPR system vendors can support workflow management applications on the same device as its physical access applications.

MEMBERSHIPS

It is the verification of membership identity that grants access to member privileges. Biometric systems are now employed in frequent traveler and frequent shopper programs, in shops and spas, at fitness centers, for pharmacy drug fulfillment, and at universities. They are used to identify special guests at restaurants and to verify member identity in drive-up lanes or self-service kiosks for credit unions or savings banks. This section will explore biometric use in retail loyalty programs, at resorts and cruises, and at educational institutions; these are just some of the growing membership venues.

Loyalty

Loyalty schemes are immensely popular to consumers. As their name suggests, these programs help ascribe much greater consumer loyalty to a particular brand. Customer loyalty marketing has become an important pillar of corporate strategy. In an era of intense competition and declining growth, customer loyalty programs are highly effective ways to stave off competition, differentiate a brand, increase revenues, build market share, and improve customer retention. It is a huge marketing advantage to cultivate customers who are not merely satisfied with but are committed to one's brand.

The key to a successful customer loyalty program is to identify and consistently reward loyal customers. Biometric solutions ease many of the hassles associated with customer identity verification. They enable loyalty points to be accumulated electronically and effortlessly. Awards can be automated. Here are some examples of how loyalty programs can be enhanced by biometric technology, especially in the hospitality industry:

- Paying for food and drinks poolside at a resort or other hospitality venue
- Rewarding customers at fast-food restaurants with points for food purchases and sweepstakes (instant rewards based on a pattern of purchases)
- Providing keyless entry to rooms or health spas
- Purchasing credits at a slot machine

Most traditional loyalty schemes cannot operate across a range of different identification methods and data recording systems. Some programs can link to a specific payment card, but few can provide multiple program access. Biometrics can do all that with a single identifier with relative ease.

Biometric-based loyalty programs can assist in targeting customers and modifying their buying behaviors. Retailers could offer special incentives to retain their best customers or tailor rewards to improve their business relationships with other customers. By recognizing customers at the point of sale, a biometric-based loyalty program can empower retailers to implement electronic marketing in real time by providing their customers with instant gratification. The customers' points are immediately updated, coupons immediately redeemed, discounts provided on the spot, and instant sweepstakes might be offered to add even more excitement to the privilege of being a member of the loyalty program—all heightening the buyer's shopping experience.

Hospitality

Nightclubs, resorts, and cruises are a key subset of the hospitality industry, and they make frequent use of biometric applications. But there is a twist here: at these venues, customers do not need to carry debit or credit cards or card tokens of any type. Whether spending one's day shopping aboard ship or enjoying the amenities of a luxury spa resort clothed only in a swimsuit, customers need only to present their hands or fingers, allow their vein patterns to be scanned, and their authenticated identities enable them to enjoy the privileges of guest membership.

Cruises generally have a lot to offer while on ship, but most require some identity authentication for billing to one's room or another means for payment for the extras, including shopping in the duty-free stores, gambling in the casinos, attending movies or night club acts, specialty drinks on the open deck, or a visit to the ship's masseuse. Since biometric solutions can offer both convenience and security, a number of cruise lines are moving rapidly to biometric identifiers for its passengers.

At ski resorts biometric systems are not needed as much for security as they are for increased accountability and convenience. Biometrics can eliminate "buddy check-ins," remove the need for fumbling for payment cards while wearing ski gear, and simply account for oneself at each stage in the skiing process.

On some family vacations the biometric focus shifts to the children. Vacation resort companies such as Club Med support kids clubs, and these resorts tend to offer a diverse set of programs. The resorts want to ensure that each child is granted access to different programs in accordance to their capabilities and needs. A biometric authentication makes that task infinitely easier and reduces the risk of the system confusing children (such as twins).

Biometrics will continue to expand its role in the hospitality industry and will especially be evident in the specialty areas such as nightclubs, resorts, cruises, sporting venues, and amusement theme parks.

Education

Back in 1972 the University of Georgia experimented with biometrics at its student dining halls to identify students for its simplified meal plan programs. Biometric technology has since caught on at many other universities as well as schools from kindergarten level to 12th grade. Today, biometrics is much more cost effective and accurate than they were in 1972, but the two primary drivers—convenience and security—remain immutable. Children do not have to remember passwords or carry a meal card. Additionally, campuses using biometrics have eliminated "buddy substitution" and identity theft at mealtime. Some students in public schools are on government subsidies for their lunches, and biometrics does not distinguish which students are buying lunches from their cash accounts and which are using government aid. The stigma of a government aid payment goes away.

School districts just want to get nutritious meals distributed to students in a timely manner. Many have tried card tokens and PINS, but children lose the former and forget the latter. Student IDs just do not work. When children forget or lose their cards, most systems require manual input by the school cashiers. Given the slow process to input student ID numbers, notwithstanding keying errors and poor memories, it would not be uncommon for lines to back up and frustration levels to rise. But a child cannot forget his hand, and VPR technology readily works with young children with relative ease. The lines into cafeterias flow smoother since students will no longer discover that they left their lunch cards at home or forgot their PINs. In Paisley, Scotland (near Glasgow), Todholm Primary School uses Fujitsu's palm vein identifier to support its existing school food catering system. The palm vein identifier is used to support children's participation in the Hungry for Success program, which ensures high quality nutritious meals for school children.

Considered a predictor of academic success at the graduate level, the Graduate Management Admissions Test® (GMAT) now requires palm vein authentication at a growing number of test centers. Pearson VUE is a global provider of electronic testing for regulatory and certification boards as well as GMAT test delivery. Pearson VUE uses the Fujitsu PalmSecure system to verify the identities of the exam takers to maintain the integrity of the GMAT process and to ensure the reliability of the exam results. Pearson VUE labors to safeguard all aspects of the exam program, including thwarting of proxy test takers who take the GMAT exams on behalf of real candidates. Pearson VUE has introduced the PalmSecure solution at 16 testing facilities. It plans to extend its biometric validation of candidates to 400 testing facilities across 107 countries.

In addition to cafeterias, biometric technology can work equally well in school libraries or for classroom attendance. This becomes more pronounced for older children who frequently change classrooms. In high school, students may need to check out supplies for laboratory, sports equipment, or other specialized items. Universities, of course, are traditional test beds for new technologies. Biometrics can facilitate physical access to dormitories and labs or logical access to university sites for assignments, submission of papers, and posting of grades.

ACCOUNTABILITY

With support from biometric systems, new standards for accountability are surfacing when boarding an aircraft, train, or ship; signing for a piece of equipment or a classified document; or recording items in a chain of evidence. Biometrics has been introduced into passports and visas in Europe, North America, and Asia. The trend is toward biometrically enabled ID cards (e.g., national ID cards, driver's licenses, health cards).

Biometrics can help make things work faster, more efficiently, and more reliably. From eliminating cash drawer pilferage to transfer of custodial property such as legal evidence, biometrics is providing accountability solutions. Accountability includes an acknowledgment and assumption of responsibility for actions and obligations incurred as a free citizen. Some sample applications include management of firearms, voting, travel, and border crossing. Each of these is discussed in the following sections.

Firearms

The focus on firearms is control—controlling access so that only appropriate individuals can activate and fire them. There are two critical applications for firearm access control: (1) privately owned firearms (POFs) used for hunting, targeting, and protection; and (2) government-controlled weaponry for the armed forces. With POFs, a biometric application can render the firearm unusable unless it is in the control of its owner or other designated users. There are over 300 million firearms in the United States alone, representing a very large firearms market—both for original manufacturing and an aftermarket where rifles and handguns are refurbished and retrofitted. Biometric readers can be fitted on rifles, shotguns, or handgun stocks and connected to the trigger housings. Indeed, a number of companies are already experimenting with fingerprint technology for their customers. The miniature fingerprint readers provide an adequate form factor for a rifle retrofit. However, fingerprint technology still has poor resistance to spoofing and is susceptible to dirt and perspiration. VPR systems are just now reaching the small size where the reader can fit into a rifle stock, and for many applications, they provide a superior solution to fingerprints.

The other firearm scenario that is biometrically addressable concerns the military. The focus here is not the individual access to handguns and rifles because the military wants all its soldiers to be able to use any firearm on the battlefield. Moreover, there are limited concerns about handguns and rifles falling into enemy hands. However, man-portable antitank (e.g., Javelins and rocket-propelled grenades [RPGs]) and antiaircraft (e.g., Stinger missiles) weaponry pose a huge concern. If these weapon systems fall into the wrong hands, they could shoot down commercial airliners and helicopters of all types, as well as destroy most vehicles, armored or otherwise.

A government's military requires multiple soldiers in a combat unit to use a given set of weapon systems in time of war, but beyond the designated and trained individuals, no one else should be enabled to use them. Of course, one way to limit access to the select few is to disable the weapon system until activated by appropriate biometric identifiers of the assigned users. In that situation, the weapon can function appropriately in combat in the hands of authorized individuals but not function when it is stolen from an armament depot or is captured on a battlefield. The problem is which biometric to use. A biometric system's form factors—usability, accuracy, reliability, and resistance to spoofing—all play important roles.

The form factors required of weapon systems on the battlefield would probably limit the selection to fingerprinting or VPR systems. Speaker recognition, eye biometrics, facial recognition, and hand geometry are all impractical solutions. Fingerprint and VPR both easily pass the form factor hurdle. However, fingerprint modalities have difficulty with the remaining criteria. The military values flexibility, and it requires a biometric identifier that has a high usability. With usability rates eschewing 2% to 5% of a given population, fingerprint systems are at a disadvantage. Additionally, battlefields run rampant with dust, dirt, and moisture, none of which are conducive to fingerprinting, as those conditions limit the fingerprint modality's accuracy and reliability. Finally, fingerprinting has a poor record regarding resistance to circumvention. Many enemies would not be averse to cutting off fingers of dead soldiers (or even live ones) to gain access to these systems (albeit liveness detection could be used to counter this limitation). On the other hand, VPR systems are more impervious to dust, dirt, and moisture; are highly accurate and reliable, even in battlefield conditions; and strongly resist spoofing.

Voting

Biometric verification of voter identity just seems like a natural application. Rather than rely on one's paper-based voter card and driver's license, currently the most popular verification means in the United States, a potential voter could be verified by vein pattern recognition.

There are many e-government initiatives that are security sensitive, none more so than electronic voting, or e-voting. Starting with a manual system of placing names in a box, devices such as mechanical voting booths, punch cards, and optical scanning machines have all been used to improve the voting process. However, technology has a way of adding additional steps to the voting process, and this is slowly attenuating the effort involved in voting. To improve speed and convenience, e-voting is now being piloted across Europe and other locations with various levels of success. Biometric verification is already in use in such diverse regions as Nigeria, Zambia, and Mexico. Some systems enable voters to use their own personal computers, entering secret codes received through the mail. Other e-voting systems require voters to use custom voting booths along with tokens containing voter-specific information.

There are several criteria that e-voting must achieve to prove viability:

- Verify the identity of each voter, but maintain anonymity of the vote.
- Properly count and safeguard each vote.
- Provide scalability while ensuring process speed.
- Prevent multiple or other unauthorized voting.

Some see e-voting as an inevitable future for e-government. However, even the most ardent supporters of e-voting are cognizant of its vulnerability to hacking and its potential threats to privacy. The essential requirement to authenticate user identity and guarantee only one vote per person while maintaining voter anonymity cannot be accomplished with today's Internet technology without including today's biometric technology.

Drivers' Licenses

There are tens of thousands of fraudulent driver's licenses. Some drivers obtain fraudulent licenses to conceal past or current illegal activities, or traffic violations in another state or in another country. Moreover, driver's licenses have become de facto identification cards in some places. In the United States, state governments are taking measures to prevent multiple licenses to the same person, an environment that nurtures identity theft. A potential solution lies in the use of biometrics.

The REAL ID Act,* passed by the U.S. Congress in 2005, calls for the establishment of national standards for state-issued driver's licenses and nondriver's identification cards. It requires states to participate in a national data-sharing program with other states when issuing driver's licenses, rendering those licenses status not unlike a national ID card. The key provision to share the data with other states is a clear attempt to curtail the issuance of multiple licenses to individuals. Although this act does not require any biometric identifiers, it does not prohibit them. Digital photos will be placed on the driver's licenses. The original deadline for implementation has been pushed back to December 31, 2009, in part because of public opposition; and it may be pushed even further out. Unfortunately, this bill was passed without national debate, and it has clear privacy implications. Four years after the passage of the controversial REAL ID Act, the Department of Homeland Security still has not released final regulations. There is much speculation that this bill will be repealed or replaced.

Other countries have chosen to move forward with similar provisions for their driver's licenses as identification cards. In many countries, there is widespread opposition to a national identification card, and at the focal point for most of these concerns are privacy issues. Whether the United States moves forward with the provisions established in the REAL ID Act, it is clear that its states require the means to accurately identify individuals, which is difficult to do without biometric identifiers.

Travel and Border Crossings

Airports and seaports present a key opportunity for identity verification. There are a number of SPT (Simplifying Passenger Travel) projects underway throughout the world. Biometric techniques are used to automate check-in, to help simplify relatively complex procedures, and sometimes to consolidate the various checkpoints, which passengers must negotiate when arriving and departing.

For reasons of government compliance, Scandinavian Airlines has implemented a biometric system that ties a passenger to his baggage at

* The REAL ID Act of 2005 requires people entering federal buildings, boarding airplanes, or opening bank accounts to present identification that has met certain security and authentication standards. The act is part of a larger act of the U.S. Congress titled Emergency Supplemental Appropriations Act for Defense, the Global War on Terror, and Tsunami Relief, 2005.

all airports in Sweden. Swedish law requires verification that a passenger with baggage boards a plane, so when a passenger checks his baggage, he must provide a biometric identifier, which is then linked to the baggage. Later, as the passenger boards the plane, he uses his biometric identifier again, and the biometric system verifies that both the baggage and passenger are on the plane prior to takeoff.

There are a number of trusted traveler schemes in trial use today. A trusted traveler scheme is one whereby a traveler who frequents a given country submits to a risk assessment background check to become registered by that country as a low-risk individual. The traveler's reward for submission to the prescreening is to avoid long lines by expediting the security check-in process (e.g., the trip-by-trip screening process) without impacting the airport's overall security.

The U.S. Transportation Security Administration (TSA) and private industry developed the Registered Traveler program. It provides expedited security screening for passengers who volunteer to undergo a TSA-conducted security threat assessment (STA) to confirm that they do not pose or are not suspected of posing a threat to transportation or national security. TSA is responsible for setting program standards and for conducting the security threat assessment, physical screening at TSA checkpoints, and certain forms of oversight. The private sector is responsible for enrollment, verification, and related services. Currently, only a relatively small set of U.S. airports and airlines participate in this program. Unfortunately, the current structure of the Registered Traveler program is more gimmick than real value as participants merely obtain the right to go to the front of the security line but still participate in the same physical screening as everyone else. However, this program has the potential to offer much more.

US-VISIT (United States Visitor and Immigration Status Indicator Technology) is an immigration and border management system that involves the collection and analysis of biometric data from visitors to the United States. The biometric data is checked against a database that identifies known terrorists or criminals. Initially, only foreign visitors who required a visa were included in the US-VISIT program; however, visitors eligible for the Visa Waiver Program have also been required to use the US-VISIT program since October 2005. Privacy issues and the 2008 elections have delayed this indefinitely.

Narita (Japan) International Airport experimented with various biometric modalities for e-check-in trials in 2002 and 2003. In 2004, Narita

Airport coordinated with South Korea's Inchon Airport for additional trials, and then the Japanese government added integrated circuit chips in selected passports for e-passport trials.

In late 2007 Japan implemented its own US-VISIT–style biometric system. Japan's revised Immigration Control and Refugee Recognition Law went into affect on November 2007, requiring all non-Japanese adults visiting Japan to provide biometric data for screening.

EMBEDDED BIOMETRICS

Biometrics is being integrated into a variety of both mobile and stationary platforms. What enables VPR systems to thrive in a variety of forms factors is the miniaturization of the image sensor. Small, highly sensitive charge-coupled device (CCD) cameras are widely attainable and relatively inexpensive. Additionally, the performance of microcomputer chips is so advanced that it can readily execute complicated algorithms. Figure 5.10 depicts a finger vein reader embedded in an ATM. VPR modules have

FIGURE 5.10 ATM equipped with a finger vein reader. (Courtesy of Hitachi-Omron Terminal Solutions.)

also been embedded in kiosks and turnstiles to improve security and efficiency.

It is the miniature form factor of the image sensor and its powerful processing capabilities that enables the VPR reader modules to be embedded in a variety of personal, multifunctional, and portable IT devices such as cell phones, personal digital assistants (PDAs), and laptop computers, significantly expanding potential VPR applications toward mobile security. Future offerings will probably include such things as keyless ignition, elevator control, and goods management. It is a form of Moore's law* that sensors will continually become smaller, algorithms will improve and become more efficient, and processing speed will accelerate. Biometric systems will continue keeping pace with the devices and applications they are infiltrating.

* Gordon Moore, cofounder of Intel, postulated in the 1960s that computer processing power will double every 18 months, at a constant level of cost. It has proven accurate for almost 50 years.

6

Evaluation and Protection of Vein Pattern Recognition Systems

True performance of a biometric system must be independently determined through a series of formal and controlled tests. Many biometric vendors claim to have a false acceptance rate that purports exceptionally low numbers, but they often do so without a third-party certification of that claim as derived by appropriate statistical tests. Such claims sometimes lack validity. By the same token, the actual performance of a biometric system is based on a set of many diverse factors from the operational environment to the cooperation of the users.

Many biometric evaluations focus on the accuracies of the algorithms used or on the engineering of the image sensors. Indeed, these are important performance measurements. However, a more comprehensive assessment of a biometric system would evaluate the matching algorithm's accuracy, processing speed, throughput rate, enrollment time, and usability and acceptability, as well as the system's performance in both a mock environment and live testing.

The effectiveness of a given biometric system for a specific application is dependent primarily on the biometric modality that is used and the application requirements that must be met. However, if the overall biometric system is not well integrated and secure, then it may be vulnerable.

This chapter describes the metrics used in performance evaluations, including enrollment, verification, and identification metrics, as well as the categories of biometric evaluations. It then describes the vulnerabilities of biometric systems to various attacks, and it shares some solutions to counter attempts at circumventing biometrics.

BIOMETRIC PERFORMANCE METRICS

There is an array of variegated performance metrics that can provide a quantitative assessment of the speed, accuracy, and other characteristics of a given biometric system. Measuring biometric performance is challenging because it requires testing with many diverse users in multiple environments over extended periods of time. Unfortunately, there is no single metric that indicates how well a biometric system will perform for a specific application. Analyses of multiple metrics may be necessary to determine the strengths and weaknesses of each biometric modality and specifically the vendor's system. A descriptive sample of commonly used metrics will be introduced as we discuss generic verification and identification applications.

Most organizations that employ biometric systems perform their own metrics measurements, and the effort required can be very illuminating, especially in gaining an understanding of how a specific biometric system functions in its environment and in tracking the system on an ongoing basis. However, it is often helpful to review comparative third-party testing where a given biometric system undergoes a full comparative evaluation in a controlled environment. The purpose of such tests is to evaluate the usability and accuracy of full biometric systems. There are several reputable companies that perform comparative biometric testing, referred to as CBT. The first major independent test inclusive of vein recognition technologies occurred in the summer/fall of 2006 when New York-based International Biometric Group (IBG) LLC evaluated Hitachi's finger vein (universal serial bus [USB] reader) solution, Fujitsu's palm vein (PalmSecure) solution, and IrisGuard's iris pattern system (Buckinghamshire, United Kingdom) in its CBT Round 6, and published its findings in a public report. IBG's CBT evaluates accuracy and usability of products using scenario-based testing. Usability was addressed by measuring failure to enroll (FTE) rates, failure to acquire (FTA) rates, and transaction duration. The authentication accuracy was addressed by measuring the false match rate (FMR) and the false nonmatch rate (FNMR).

The approximately 650 test subjects were New York City residents, and they exhibited a wide range of ethnicity and age. The comparison of the results among the iris recognition system and the VPR systems were more favorable toward the VPR systems.

CBT is one of several highly reliable third-party testing methods. IBG has evaluated more than 50 biometric systems, and its results have a wide distribution. The CBT methodology is compliant with ANSI* INCITS† 409.3-2005, and the scenario testing is compliant with ISO‡/IEC§ JTC1SC37 19795-2: Testing Methodologies for Technology and Scenario Evaluation. Throughout this chapter, some of the findings of that evaluation are cited.

Enrollment Metrics

We have discussed how a good biometric is universally usable for and acceptable to a population of end users. Therefore, one measurement of usability occurs during enrollment and applies to both identification and verification applications. Recall that enrollment is a critical stage where people must provide a biometric characteristic so that a reference template can be created and stored for future matching. However, sometimes people are unable to enroll. This is an essential measurement for assessing large-scale biometric systems.

A biometric system's failure to enroll (FTE) is the rate at which potential users are rejected from enrollment in a biometric system due to insufficiently distinctive biometric samples or their inability to interact with the system. For example, an individual with a musculoskeletal disorder might be unable to provide a fingerprint image because he cannot place his finger on the appropriate sensor. In any population of users, there will be individuals whose physical attributes militate against successful enrollments.

* ANSI is the American National Standards Institute. It is a private, not-for-profit organization that oversees the development of voluntary consensus standards for products, services, processes, and systems in the United States. ANSI coordinates American standards with international standards.

† INCITS is the International Committee for Information Technology Standards. It is accredited by and operates under rules approved by the ANSI to ensure that voluntary standards are developed by the consensus of directly and materially affected interests.

‡ ISO is the International Organization for Standardization. (Note: The order of the letters in the acronym comes from the original French name.)

§ IEC is the International Electrotechnical Commission. It manages conformity assessment systems, which certify that biometric systems and their components meet its international standards.

Another example is that a professional brick layer might not have distinctive fingerprints. A person who lacks keyboarding skills would have difficulty enrolling in a biometric system based on keystroke dynamics. Women in some cultures may not wish to remove veils for facial recognition. Children may not follow instructions well. Some users simply are so nervous or high strung that they cannot stop fidgeting even for a few seconds to use an iris recognition scanner or to participate in a speaker recognition system. Other reasons for FTE include inadequate user training, poor supervisor training, device time-outs, or the inability of a system to sense or locate a presented sample. FTE figures includes those who, for physical or behavioral reasons, are unable to present the required biometric feature. Thus, the FTE rate is the proportion of users for which a biometric system cannot generate acceptable reference templates. To properly address these situations, there must be a contingency plan to accommodate individuals who are unable to satisfactorily complete an enrollment.

There are several advantages of VPR systems relative to enrollment that reduce FTE rates. First, VPR systems tend to be intuitive and very user friendly. Second, since veins are located beneath the skin, they are not prone to external distortions like some physical characteristics. Third, VPR systems exhibit an exceptionally high usability rate in that most people (approximately 98.8%) have scanable veins in their fingers and hands. Last, VPR systems can use multiple instances (e.g., fingers and hands) for enrollment. Most VPR systems recommend at least two instances be enrolled. A given instance should not be declared unable to enroll until at least two failed enrollment attempts, and most systems allow more than two. If none of the instances selected can be enrolled, then the individual would qualify as an FTE.

Enrollment results including transaction durations are generally functions of the biometric hardware, the algorithms used, the quality of the instruction and training, and the deployer's selected decision policies. Operational biometric systems may achieve higher or lower FTE rates, and transaction durations may vary based on customized implementations. Of all metrics, enrollment duration is the most dependent on an application's constraints. Examples include the need to hurry an enrollment for a user in a retail check-out line while attenuating the same process during an employee enrollment to enable additional training. Transaction duration should include time for the individual to align himself with the vein pattern reader, time for the device to locate the instance (usually, this takes only milliseconds), and time for the system algorithms to validate the quality of the sample (also usually measured in milliseconds). Since most vein pattern systems use an alignment mechanism to quickly position the hand or finger and since the use of a vein reader is fairly intuitive, enrollment in a VPR system tends to require less time than some other modalities. During CBT Round 6 IBG testing, the medians for the enrollment of Hitachi and Fujitsu VPR systems were 33.3 and 61.7 seconds, respectively. For verification transactions, the median durations were 1.23 seconds and 1.77 seconds, respectively.* The IBG report stated, "Hitachi and Fujitsu achieved FTE rates that can be considered exceptionally low for this type of testing. Indeed, both systems failed to enroll only a single transaction out of nearly 1300."† That is a failure rate of only .08%; and it validates VPR as a high usability biometric modality.

Verification Metrics

As we have discussed in previous chapters, verification is the authentication of someone's identity claim. With biometric systems in a verification mode, a user is making a claim regarding a specific identity. The biometric system assesses that claim by attempting to match the sample biometric template of the user to the stored reference template that was produced at enrollment. The biometric system produces a probability score regarding the claim. That score is compared to a specific system threshold, and a decision is rendered. For measurement purposes, it might be prudent to test at three thresholds: low, medium, and high security.

In Figure 6.1, a user approaches a biometric system and claims to be Jose Gonzalez. He submits his hand to the VPR reader, which creates a

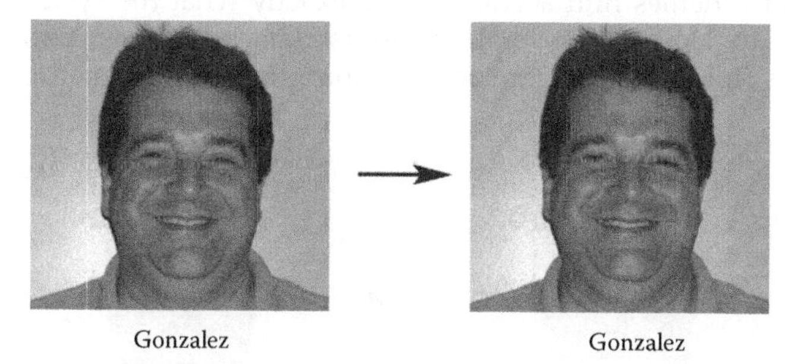

FIGURE 6.1 1:1 match.

* IBG CBT Round 6 Public Report, September 2006, p. 5.

† IBG, p. 41.

template to match against the stored template from enrollment. The verification threshold is set at 95, and Mr. Gonzalez's biometric score is 97.5 and deemed a match. We can safely say that Mr. Gonzalez is who he says he is.

One of the metrics used for verification systems is a false rejection rate (FRR), which measures the portion of genuine users that are rejected. It measures the percentage of times the biometric system produces a false reject (i.e., an unintended reject). This occurs when an authorized individual is not correctly matched to his or her existing reference template. For a verification system, it can be estimated as

$$\frac{\text{Number of False Rejects}}{\text{Number of Verification Attempts}}.$$

For example, Mr. Gonzalez has cause to use this biometric system daily, and on one particular day the VPR system calculates a similarity score for Mr. Gonzalez at 93.7. When the calculated similarity score is compared to the verification threshold, Mr. Gonzalez is denied access. Since Mr. Gonzalez really is Mr. Gonzalez and is not an imposter, the biometric system has erred and produced a false rejection. As Mr. Gonzalez and his business colleagues continue to use the biometric system, we would see over time how often anyone received a false rejection.

In Figure 6.2, we see someone else claim to be Jose Gonzalez. In this case, the imposter submits his vein pattern identifier for evaluation, and the system calculates a similarity score of 62, well below the threshold of 95; and it denies him access. This is exactly what the system is supposed to do. However, several imposters claim Mr. Gonzalez's identity numerous times to try to gain access, submitting their biometric samples

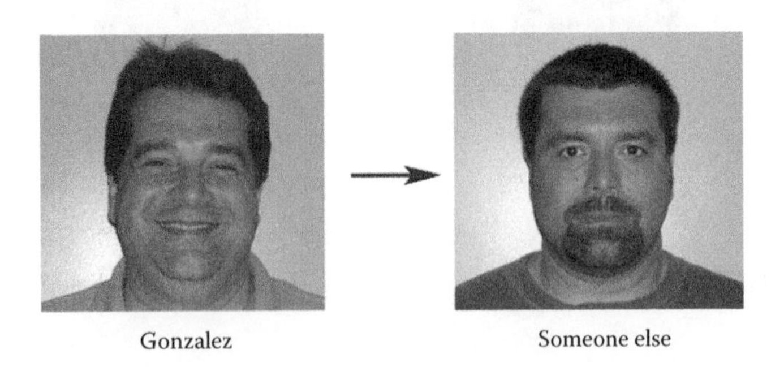

FIGURE 6.2 1:1 nonmatch.

over and over again. Eventually one of them is successful and gains entry. The biometric system scores him a 95.5 and agrees that the imposter is Mr. Gonzalez; but of course, he is not. This is called a false acceptance. And if we gather information regarding how often the biometric system had approved a false claim, we could calculate a false acceptance rate (FAR), which measures the percentage of times the biometric system accepts an imposter. A false acceptance happens when a sample template from an unauthorized individual is incorrectly matched to a reference template of an authorized individual. This rate can be restated as a probability that a biometric system will incorrectly verify an individual or will fail to reject an imposter. For a verification system, it can be estimated as

$$\frac{\text{Number of False Acceptances}}{\text{Number of Imposter Verification Attempts}}.$$

Failure to acquire (FTA) is similar to FTE except it occurs during verification transactions as opposed to enrollment transactions. FTA is the rate at which a biometric system fails to capture or extract usable information from a biometric sample. *Usable*, as used here, refers to an image or signal of sufficient quality to enable a comparison. This could occur for a whole range of reasons including equipment malfunction, interference with the image sensor (e.g., dirt particles on the camera), environmental issues, or human anomalies such as improper alignment of the user's finger or hand, or even adjustable thresholds for image or signal quality. Generally, when a biometric system allows multiple attempts, FTA measures the biometric system's failure to capture throughout these multiple attempts.

The IBG evaluation's public CBT Round 6 report stated, "Fujitsu and Hitachi were able to capture a very high percentage of hands and fingers, respectively ... The low failure to enroll and failure to acquire rates generated ... were surprising."* Although the CBT evaluations of VPR technology involved only Hitachi and Fujitsu, it is quite probable that other VPR vendors would have received relatively low FTE and FTA rates.

Authentication accuracy is an important determinant for biometric systems. Defining and achieving accuracy is a prerequisite for assuring security of the biometric system. Moreover, that accuracy must be repeatable and comparable at multiple independent tests and sustainable across the life of the biometric product. Biometric system designers can accommodate a desired level of accuracy by setting the verification threshold to

* IBG, p. 10.

whatever is deemed an appropriate level. Most biometric systems have the capability of making threshold adjustments.

As mentioned at the beginning of this section, performance verification testing at each of three widely differing thresholds would give a good indication of the affect the threshold settings have on the FARs or FRRs. Theoretically, it would be great to raise the verification threshold of the biometric system so that it is very difficult for an imposter to receive a similarity score higher than the threshold. The sensitivity of the system might be set so high as to require near perfect matches of reference and live data. This most probably would reduce the FAR significantly, but it would most likely increase the FRR considerably as well. That is because the FAR and FRR are mathematically connected. On the other hand, lowering the verification threshold such that the reference and sample template only approximate each other would result in less false rejections but potentially higher false acceptances. Enterprises must determine an acceptable compromise level and tune the system sensitivity to the desired level of accuracy. A biometric system can only operate on one threshold setting at a time. Therefore, adjustments for false rejections or false acceptances are generally made with respect to the application supported and the security level needed.

When a biometric system provider purports accuracy levels such as a FAR of .00001% and a FRR of .01%, the claim does not mean that the biometric system can achieve both accuracy levels simultaneously. What the biometric vendor is advertising is that its system can achieve either of these levels, depending on the threshold setting. Indeed, this is the rationale of many biometric vendors in offering three thresholds: one on each extreme and one in the center. Additionally, when evaluating the professed accuracy levels of a given biometric system, a potential buyer should ask the question, "Who certified these accuracy levels?" The fact is that only a small proportion of biometric vendors have submitted their system to a thorough and professional third-party review. Indeed, although most biometric providers do perform extensive and comprehensive internal testing, their testing can be subject to a variety of influences that render the results inconsistent when scrutinized. There are simply too many variable conditions that can affect testing results; and thus, standard, independent tests in controlled conditions are preferred for benchmarking. For truly cross-comparative and accurate results, a highly reputable third-party certification authority should verify the biometric system's accuracy figures.

If we were to graphically plot the FAR and the FRR, as was done in Figure 6.3, then we would create an error trade-off graph. Biometric system

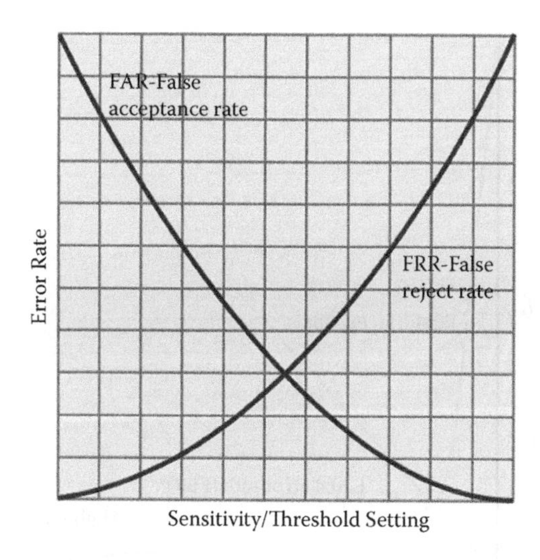

FIGURE 6.3 Error trade-off graph.

operators use this graph to help determine where to set their threshold sensitivities. A setting to the far right would mean a low FAR and a higher FRR. A setting to the far left would have the opposite result—a high FAR and a low FRR. The optimal threshold setting depends on what the biometric system operator is trying to achieve in terms of error trade-offs. Although this graph depicts threshold sensitivities, it does not enable an accurate performance chart independent of the threshold setting. That is because an error trade-off graph uses a threshold setting as one of its parameters. But this is easily remedied. We can remove the threshold dependency by plotting a FAR versus a FRR on the respective x and y axes, creating what is called a detection error trade-off (DET) curve.

A DET curve is a useful means to accurately depict system performance across a range of threshold decisions. A DET curve may plot error rates on both the x (FAR) and y (FRR) axis in a log scale for every threshold value, or it may plot its curve using a normal deviate scale. Plotting with a normal scale tends to be more linear, and its reader can more easily discern the differences among similarly performing systems. With real-world applications, the FAR and FRR are traded against each other by manipulating one or two parameters, as calculated in an equal error rate (EER), also referred to as the crossover error rate (CER). The EER is the location on the DET curve where the FAR and the FRR are equal. Figure 6.4 depicts a stylized (non-real world) DET curve depicting the FAR vis-à-vis the FRR. This

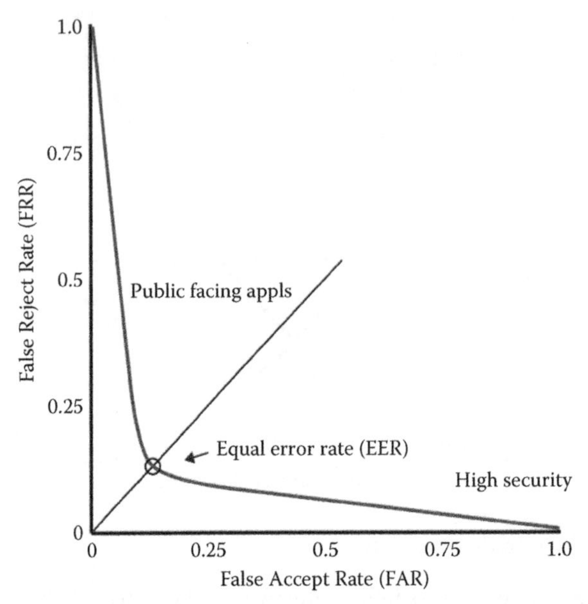

FIGURE 6.4 Detection error trade-off (DET) curve.

information helps the system administrator make system adjustments in favor of security and convenience, as applicable. In some biometric systems, such as one that is public-facing, the FRR rate might be set very low so as not to irritate legitimate users, while accepting the risk that an unauthorized person might be allowed access. FAR is to security what FRR is to convenience. For other applications such as a high security facility, the FAR might be set very low to ensure against potential intruders while tolerating the employee hassles associated with a higher FRR. In these latter situations, human intervention may be needed to remedy the false rejects.

There is a relationship between a FAR and the probability of verification. That relationship, referred to as a receiver operating characteristic (ROC) curve, is plotted in Figure 6.5. ROC curves plot the percentage of false positives vis-à-vis the percentage of true positives (e.g., verifications or identifications). A ROC curve is a comparison of two operating characteristics as the criterion alters. ROC curves are similar to DET curves except that the *y* axis for a ROC curve is the probability of verification (or identification). Biometric system providers want the area under the curve

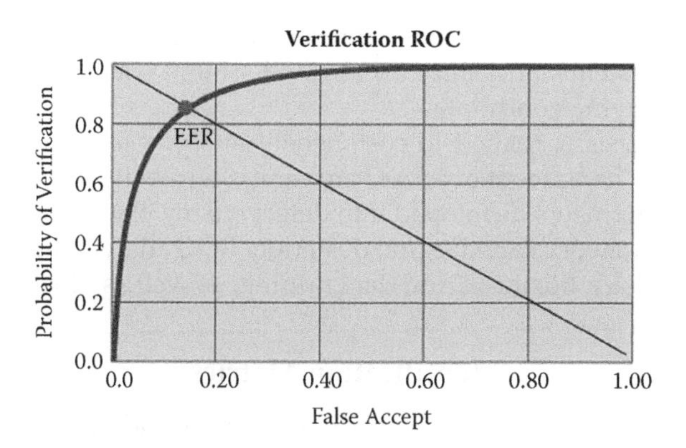

FIGURE 6.5 Verification receiver operating characterisitc (ROC) curve.

to be as large as possible. For verification applications, the ROC curve illustrates a biometric system performance by plotting the rate of false accepts on the x axis against the corresponding rate of genuine matches on the y axis. ROC curves are very useful in either comparing the performance of different systems under similar conditions or in comparing the performance of the same system under differing conditions. Thus, ROC curves are threshold independent.

The EER can be determined by drawing a straight line from the upper left corner to the lower right, and observing the point of intersection with the curve. The EER is only one statistic used to evaluate the biometric performance of an identity verification application. To some degree an EER provides an impartial measurement for use in comparing the performance of two or more biometric systems. Unfortunately, since many biometric systems are not set to operate at the EER level, its usefulness may be confined to comparing biometric system performance. Theoretically, a system's EER can be used to provide a threshold performance measure. In its CBT Round 6 evaluation, IBG calculated the transactional EER for Hitachi's FV system at .29%.* That means that if the threshold were set to the EER, then for every 1,000 verification transactions, only 3 transactions would result in a false accept or false reject response. Generally, the lower the EER, the better the system's performance, as the total error rate decreases. However, a comparison of EERs among alternative systems remains more of a hypothetical exercise than a fully accurate one

* IBG, p. 7.

since EERs are not necessarily representative of performance in most operational systems, and since each tested system uses different testing samples in varying conditions.

ROC analysis is related to cost–benefit analysis for diagnostic decision making. The concept of a ROC curve was borne during World War II when radar engineers developed it to detect enemy objects in battlefields. It grew from a theory called signal detection. ROC analysis has since been used in radiology, business, and data mining, as well as biometrics.

Identification Metrics

Synonymous to a one-to-many (1:n) comparison, identification locates an individual within a population. Identification can confirm or deny that an individual associated with a given biometric characteristic is not enrolled, or that he is or is not on a predetermined watch list. It is convenient to partition identification applications into open and closed subsets.* In an open identification application, also known as a watch list task, the biometric system determines if a given biometric sample matches the template of someone in the database. In this environment, the individual in question is unknown, and it is not known in advance if the person is even in the database. A watch list application asks, "Are you in my database; and if so, can I find you?" A watch list application could be looking for a missing person or trying to identify a dead body. The Federal Bureau of Investigation's (FBI) Integrated Automated Fingerprint Identification System (IAFIS) is actually a watch list application.

In Figure 6.6 the biometric template taken from the unknown sample is compared to each image in the database, one by one. As each comparison is made, a similarity score is generated and compared to the threshold score. If the similarity score exceeds the threshold score, then the biometric system sends a predetermined notification. A review of the information may result in a correct identification. If the threshold is set too high, it may miss a correct identification; if it is set too low, there will be multiple incorrect identifications, known as false alarms. If one were to map these occurrences over time, he could graph a watch list ROC, which would illustrate the trade-off between the false alarm rate and a correct identification rate, as shown in Figure 6.7.

* Some biometric practitioners divide all biometric applications into three sets: verification, watch list, and closed identification.

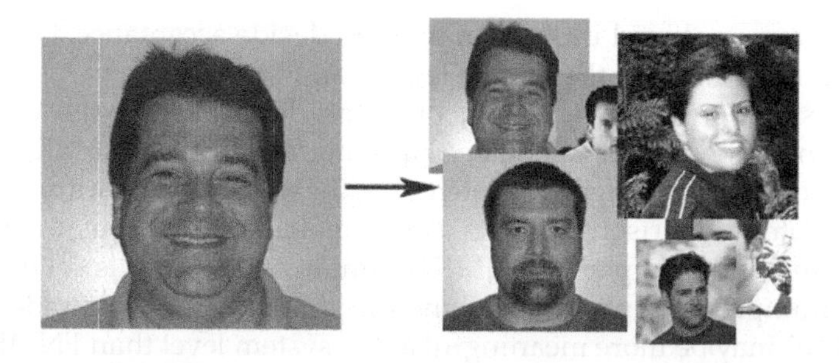

FIGURE 6.6 1:n match.

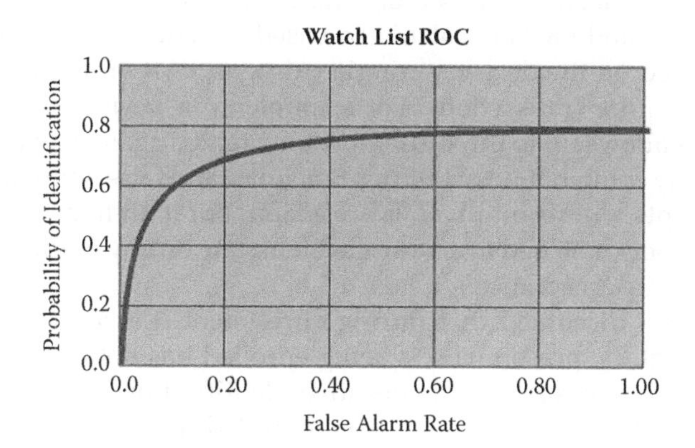

FIGURE 6.7 Watch list ROC.

There are some applications in which the concepts of acceptance and rejection may be reversed, thus turning around the meaning of false acceptance and false rejection. If an organization wanted to determine if anyone had previously enrolled in a given system, either by error or on purpose, it could run a watch list application on a given biometric sample to see if there were any matches in its existing database. Thus, if it had a number of instances in which a match did occur, that would be considered its match rate. For those instances in which they received a match notification that turned out to be incorrect, they would be able to calculate a false match rate (FMR). This is the rate of incorrect positive matches by the matching algorithm for single template comparison attempts. For those biometric systems that use just one attempt to decide acceptance, FMR is conceptually similar to a FAR. In those situations in which multiple attempts are combined in some manner to decide acceptance, FAR could be more meaningful at the system level than FMR.

For situations in which someone was duplicated in the database, the organization would calculate a false nonmatch rate (FNMR). This is the rate for incorrect negative matches by the matching algorithm for single template comparison attempts. For biometric systems that use just one attempt to decide acceptance, FNMR is similar to FRR, but it is not the same. When multiple attempts are combined in some manner to decide acceptance, FRR may be more meaningful at the system level than FNMR.

The differences between FRR/FAR and FMR/FNMR essentially reflect the differences between decision errors and matching errors. FMR and FNMR are calculated based on a number of comparisons or matches; however, FAR and FRR are calculated based on a number of transactions and are based on the decision criteria. Also, an FAR or an FRR includes failure to acquire transactions. For example, in a positive identification system that allows up to three attempts for a match against an enrollment template, any combination of failures to acquire and false nonmatches over three attempts will result in a false rejection. But if an image is acquired and then falsely matched to an enrolled template on any of three attempts then it is a false acceptance.

Chapter 3 discussed that during enrollment it is a very good idea to ensure that the person who is being enrolled has not been previously enrolled. This is referred to as negative identification. The rationale for this check is that someone may have enrolled and forgotten that they did so or may not have realized that he enrolled in a particular program; alternatively, someone may be attempting to compromise the biometric system through multiple enrollments. Negative identification is effectively a watch list application. Increasingly, governments at both national and state/province levels are using this identification technique to determine if citizens on benefit programs are enrolled under other names in other locations.

Figure 6.8 plots a closed identification ROC that compares false alarm rates to detection and identification rates. A closed loop identification application is a subset of the open loop since the biometric in question does indeed belong to an individual in the database. A closed identification application asks the question, "Can I find you in my database?" The procedure is exactly the same as for a watch list application. However, there are few applications that actually are pure closed loop. Notice that the probability of identification for a closed ROC shown in Figure 6.8 is a bit higher than the watch list ROC in Figure 6.7.

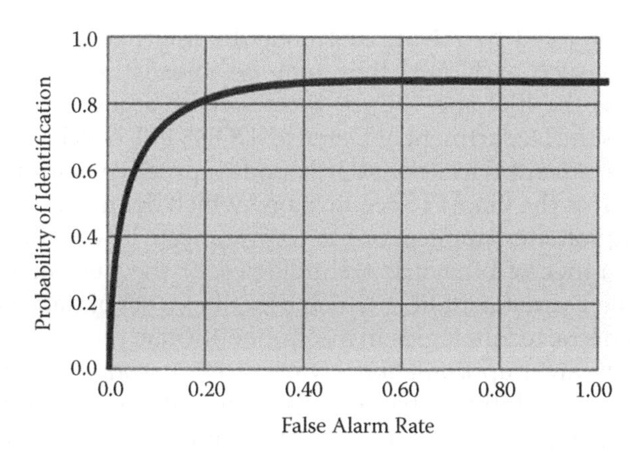

FIGURE 6.8 Closed identification ROC.

BIOMETRIC STANDARDS AND TESTING

There are numerous and variegated biometric standards, and the industry is producing more all the time. A biometric standard generally originates from some recognized government organization or consensus body, and it provides common, repeatable rules or guidelines, interfaces, or behaviors to which organizations must adhere. These standards are developed with multiple participants as they represent a consensus of individual thoughts. There are a variety of organizations that develop, approve, and maintain consensus biometric standards.

The responsibility for developing biometric product and process standards differs by country. In the United States, the responsibility lies with the private sector. The U.S. government relies on voluntary standards, which are often incorporated into its regulatory requirements and almost always into its procurement requirements. However, agencies of the U.S. government do participate in standards development and try to "harmonize" the open standards, processes, and testing methodologies.

Standards organizations may be national organizations such as ANSI or BSI,* or they may be nationally accredited, such as INCITS or NIST

* BSI is the British Standards Institute, the world's first national standards organization. It helps to develop national, international, and private standards, and it tests and certifies selected products.

ITL (both accredited by ANSI);* they may be international treaty-based organizations such as ICAO;† they may be international private-sector based such as the ISO and IEC; or they may be a government agency-based such as the Department of Defense (DOD), FBI, or NIST (agencies of the U.S. government). These standards bodies produce helpful biometric solutions such as the Bio API Specification,‡ which defines an open system API enabling software applications to communicate in a common method with a wide range of biometric technologies, or the Common Biometric Exchange File Format (CBEFF),§ which describes a set of data elements to support biometric technologies in a common format.

Standards are important for a variety of reasons. The paramount reason is interoperability of biometric systems, along with compatibility of components within a biometric system. But there are other benefits as well, including interchangeable processes, training, vendor independence, safety, reliability, efficiency, and, of course, lower prices. With standardization, vendors are channeled in a particular direction, and their competitive energies focus more on intangibles such as performance, quality, and form factors. Standards help prevent vendor lock-in; they simplify system design and enable support of multiple technologies. Biometric standards help improve the overall performance of biometric solutions; they reduce risks, and they promote the widespread use of biometrics.

Standards generally specify requirements; indeed, they often reflect the functional requirements and performance needs of a given biometric approach. They govern how biometric components should act and interact. Biometric standards are required for all aspects of a biometric system including formats, interfaces, testing methods, data quality, collection and storage, application profiles, performance metrics, and even one's biometric vocabulary. To make a proper comparison of one system to another, or to effectively evaluate a single biometric system and communicate the results, standard terminology, standard testing, and standard processes are needed. The use of standards also conveys a level of performance and

* ITL is the NIST International Technology Laboratory. It develops tests and test methods that enable biometric industry stakeholders to objectively measure, compare, and improve their biometric systems.

† ICAO is the International Civil Aviation Organization, a UN Specialized Agency, which functions as the global forum for civil aviation.

‡ Created by ANSI/INCITS 358.

§ Developed by NIST in cooperation with several external associations and standards working groups.

reliability that has liability and privacy implications, and which affects the procurement decision to acquire a biometric system.

There are many types of biometric testing that can occur. In general, the biometric industry has categorized three broad types of biometric evaluations: conformance, interoperability, and performance tests. Each has a specific area of specialized testing; but in practice, evaluation methods are often combined or conducted sequentially.

Conformance Tests

Biometric conformance assessments are basically the process of checking, via a set of test assertions, whether a biometric system reliably implements a given standard or profile. A conformance test, also called a compliance test, determines if a given biometric product has met a professed set of functional requirements as specified in the standard. The specifics of a conformance test vary with the standard(s), which a conformance test addresses. For example, API standards generally require extensive functional assessments of all "call routines," including error conditions. On the other hand, data format standards generally require a review of all inputs, outputs, and entity object relationships.

Product manufacturers, resellers, integrators, users, or independent, third-party organizations can perform conformance testing. The last category of testers is more likely to provide an unbiased evaluation regarding whether the product conforms to the application standard(s). Indeed, most compliance testing is performed by external organizations, including not only private enterprises specializing in this service, but sometimes by the standards bodies themselves. Such focus provides a certified guarantee by these external bodies that a given biometric product meets with the established standard(s). Indeed, well-documented, standards-based conformance testing imbues in developers, integrators, and users increased levels of confidence in product conformance claims.

Conformance testing can occur throughout a biometric product life cycle, especially during the development phase. A multitude and variety of test procedures have been developed specifically for conformance testing. Indeed, some organizations such as NIST are developing supporting test tools; and NIST continues to work on the development of Conformance Test Suites for biometric interface standards.*

* Conformance test suites are test software that helps determine conformance to a testing methodology described in a specification or standard. Their stated purpose is to support users and developers whose products must perform to specific biometric standards.

There is a movement in the United States spearheaded by INCITS to try to harmonize the conformance testing among various organizations. Agreements are being rendered on such things as conformance testing methodologies, test tools, testing laboratory requirements, and accreditation and certification bodies.

The results of conformance testing enable vendors to identify and diagnose potential issues early in the development process, and to make objective evaluations of their biometric products. A biometric product that is certified as compliant to a specific set of standards can improve market viability and increase product sales. Conformance test results improve the decision making of product buyers, decreases risk of procurement, and increases confidence in the overall biometric system that is implemented.

Interoperability Tests

Interoperability refers to the general ability of otherwise disparate biometric systems or components to exchange data accurately, consistently, and effectively. Once exchanged, the data must be usable. So interoperability testing validates the degree to which a biometric product can exchange usable information and interoperate with other biometric products.

Interoperability is not just a desirable benefit; it is an essential to the efficient operation of large-scale biometric systems. Interoperability testing is a biometric evaluation to determine to what degree a biometric product can interchange information and interoperate with one another or more biometric products at a given level of performance. Often test results of biometric systems are not comparable as interoperability of the testing processes or the components used are nonexistent. Understanding the degree to which a biometric system or component is interoperable with another system or component is crucial to a biometric system's implementation success.

Interoperability is related to both the standard and its implementation. Before testing for interoperability, a general prerequisite is testing for compliance to established standards. Claims of conformance and interoperability should be validated by preset test metrics and test conditions.

Performance Tests

Different biometric tasks—verification, closed identification, and open identification—require different performance characteristics. Moreover, their respective relevance and value can vary. Biometric performance evaluations test the performance of a biometric system, or its component parts, in accordance with a derived set of requirements. Performance tests are usually categorized into three types of evaluations: technology, scenario, and operational.

Technology evaluations are usually performed in laboratories, preferably by disinterested third parties. Typical evaluations are focused on the biometric systems software, rather than the hardware. The evaluations might include algorithm testing, sensor testing, or comparison testing. For example, the evaluation might compare recognition algorithms for a given modality in order to benchmark the modality's capabilities, progress, and status. The testing generally occurs with a standardized database garnered by a generic sensor for that modality. The purpose of technology evaluations is to validate the technological capabilities of a given system and to help determine technology approaches that may offer the most promise in the future. They tend to be generic, repeatable standard tests applied to a given biometric modality and type to determine the technical capability and scalability of the biometric algorithms or sensors. These types of evaluations occur with some frequency. Results may be published. ID Technology Partners, based in Gaithersburg, Maryland, is an example of an enterprise that provides biometric technology evaluations (among an array of other testing and services). NIST is an example of a government agency that performs technology evaluations. Technology evaluations provide key performance statistics, and often identify areas of the biometric system that require additional research and development focus.

Technology evaluations are often conducted in two phases: a training phase and a testing phase. In the training phase, biometric samples are collected from test subjects and they are thoroughly documented. The data is usually independent of the biometric device, although the data is representative of the samples that the biometric device would acquire. Further, the testing tends to use an off-line processing of the data. In the test phase, the testers define the performance statistics and determine how they will be estimated from the test data. The system matchers are then tested using newly collected data or the previously collected data. The computed scores will enable estimates of FMR and FNMR (or FAR and FRR for authentication).

Scenario evaluations assess the sample acquisition sensor and biometric system algorithms for a specific, simulated application. The purpose of a scenario or application test is to determine whether the selected biometric system can satisfy the acceptance criteria in terms of system performance. The results of a scenario evaluation should prove a strong indication regarding how a particular biometric system will operate in a specific environment or for a specific application. This is an end-to-end biometric evaluation in which all aspects of the system are tested. They can be third-party managed or privately managed by a given enterprise. Scenario evaluations generally take place in an environment, such as a testing facility, that can model real applications, and which strives to employ repeatable approaches, such as using written test scripts. Scenario evaluations generally use a controlled set of test users. However, the modeled test must be carefully controlled. The volunteer test subjects tend to use the tested biometric system over some defined period of time. Because volunteers tend to be eager participants, as opposed to technology-adverse subjects, care must be taken to not allow the results to be skewed. The scenario evaluation results are generally not made public,* and they enable additional system integration to take place prior to implementation. Although simulated, the scenarios attempt to address real-world applications.

Operational evaluations determine the performance of a biometric system in a specific environment with targeted users. It is usually an on-site test, sometimes conducted by the same people who will use the system. The purpose of this evaluation is not to review the technology or to perform an overall system evaluation, but it is an examination of workflow performance and impact on the organization for the specific application with a specific targeted population. These evaluations generally occur prior to formal implementations. The purpose of such operational tests is to validate the performance of a given system outside a controlled laboratory environment to determine if any unanticipated issues surface.

Operational evaluations should occur under realistic circumstances to help streamline procedures and assist the population of end users to better use their new biometric system. The results of operational evaluations enable enterprise decision makers to fine-tune their implementation plans prior to enrollment activities. Operational test results are usually not repeatable due to differences among operational environments. Measurements of performance characteristics—the FAR, FRR, FTA, and so forth—are taken during this phase.

The performance metrics of any biometric system can be estimated with confidence only when they have been tested across a large database obtained from diverse individuals under a variety of environmental

* A clear exception to this approach is IBG's Comparative Biometric Testing (CBT), which releases the results of its scenario-based testing to the public.

conditions. Further, although multiple organizations, including biometric system manufacturers, system integration, and special interest groups can and should conduct biometric evaluations, third-party evaluations are the only testing that can provide an objective review.

As stated in IBG's CBT Round 6: "In summary, vascular recognition ... appears to be a very serious competitor to fingerprint, hand geometry, and certain iris recognition systems used in large-scale 1:1 access control, logical access, and consumer ID applications. The systems tested [Fujitsu and Hitachi] provided a strong combination of usability and accuracy.*

Biometric evaluations are important for a variety of reasons. They have advanced biometric technology by citing shortcomings and limitations, and by setting the stage for improvements. Positive evaluations raise the performance bar and negative evaluations focus attention on the need to improve and usually provide some guidance on what needs improving. As a result, biometric systems in general have evolved to a level that supports a high degree of accuracy and user acceptance.

Globally, there are many organizations attempting to define criteria for the evaluation of biometric systems. However, there is not yet a generally recognized manner to evaluate biometric systems across all modalities, and to provide an unbiased comparison of their strengths and weaknesses. For now, any assessment of a given biometric modality must be limited to the context of specific applications.

CIRCUMVENTING BIOMETRICS

As long as people create security systems to safeguard information and property, there will always be individuals or organizations who will attempt to circumvent that security. Whatever biometric system is produced, it must be relatively resistant to multiple types of circumvention. A form of identity theft occurs when an imposter tried to access a system for which he is not authorized, or when the individual attempts to enroll in a biometric system more than once, creating multiple identities. These are examples of the types of fraud that a biometric system must thwart. There are multiple ways to attack biometric systems depending on the modality and the quality of the overall system.

* IBG Report, p. 10.

FIGURE 6.9 Gummy bears.

Attacks on biometric systems include spoofing,* replay attacks, and template database attacks. Some biometric systems have been defeated by submission of fake hands and fingers (with lifted fingerprints) or facial photography. Cryptographer and professor Tsutomo Matsumoto of Yokohama National University gained international notoriety demonstrating how relatively easy it was to spoof some fingerprint systems using "gummy fingers,"† which are created using gummy bears (as shown in Figure 6.9) as the gelatin base.

Spoofing requires some basic knowledge of the targeted biometric modality, and sometimes it needs some level of collusion either with enrollment operators or with the targeted "victim," referred to as an insider attack. There can also be sabotage issues during the enrollment process. By purposely providing poor quality or false images at enrollment, individuals can weaken the integrity of the biometric system. If enough individuals submit unclear samples, then the biometric system might be rendered ineffective.

* Spoofing is the process of defeating a biometric system through the introduction of fake biometric samples. This is a somewhat narrower definition than some biometric writers employ. For example, synthetic fingerprints can be created on the surface of a variety of materials, including gelatin, latex, and silicon.

† Tsutomo Matsumoto's paper "The Impact of Artificial 'Gummy' Fingers on Fingerprint Systems" (*Proceedings of SPIE*, Vol. 4677, Optical Security and Counterfeit Deterrence Techniques IV, January 2002) became famous as the inventive professor traveled to international conferences and demonstrated how he could lift a fingerprint from a pane of glass and overlay it on a gelatin substance resembling a finger (as found in gummy bears) using an electron microscope, an inkjet printer, and Photoshop software.

In some cases inhibitors to verification or identification are not caused by fraudsters, but by the biometric system itself or by external factors. However, fraudsters often exploit these inhibitors. A variety of potential issues can occur, including noise (such as defective sensors, unfavorable ambient conditions) or intraclass variation (where biometric data acquired during authentication differs from the reference data), can inadvertently leading to false rejects or false acceptances.

Additionally, there are some physical concerns. For example, an unattended biometric device might be vulnerable to a physical attack if there were a way for the attacker to open the device to defeat it. For this reason, many biometric readers now include tamper-resistant countermeasures such as special locking devices or alarms. Those systems that have networked biometric sensors that transmit to various processors might be intercepted. That is why all biometric data transmissions must be encrypted. In addition to thwarting efforts to fool a given biometric, there must be safeguards in the system itself to guard against playback attacks and other electronic attack techniques.

Yet another method to undermine a biometric system is to attack the other components of the overall security system. Security is only as strong as its weakest link, and the biometric component can be rendered ineffective if fraudsters can work around the biometric system. A good example of this is the high incident of "tailgating" that occurs at some enterprises that use biometric access. That is, someone who has gained entry to a building or office by authenticating himself holds the door open for others who bypass the biometric reader. A corollary to this is exception handling, which may introduce a security weakness as procedures to work around the biometric system become institutionalized in the enterprise over time.

Attacking a Biometric System

Susceptibility to attacks is a key concern regarding any biometric system, and *all* biometric systems have vulnerabilities. Detecting and addressing biometric fraud is challenging, but thwarting attacks on biometric systems is critical to maintain the confidence of the system stakeholders. Antispoofing techniques can significantly raise the level of difficulty for fraudsters. But spoofing the biometric sensor is only one of eight types of potential attacks on biometric systems, albeit spoofing constitutes the majority of biometric circumvention efforts. As depicted in Figure 6.10, the following are generic vulnerability points in a biometric system:

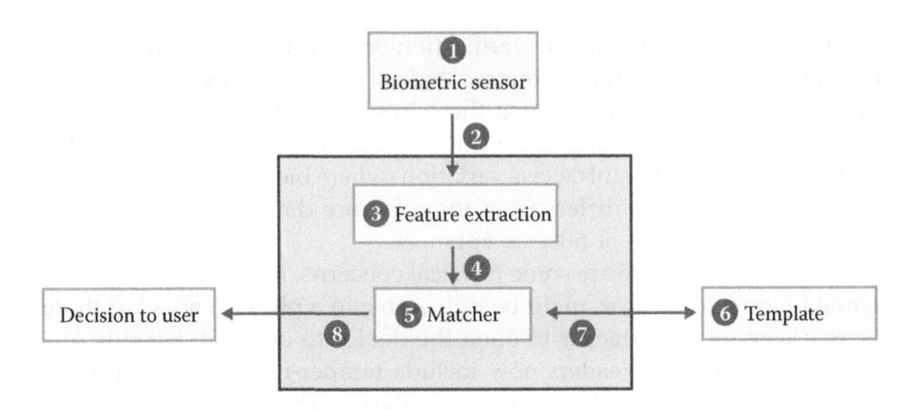

FIGURE 6.10 Biometric system attack points.

1. *Attacking the sensor*—Fake characteristics or artificial traits such as gummy fingers or other fake images, can fool some biometric sensors. This practice is also referred to as spoofing. In a spoofing scenario, an attacker creates a fake copy of a valid biometric identifier to deceive the sensor in order to gain entry.
2. *Replay attack*—A reproduced digitized biometric signal or image is replayed to the biometric system, bypassing the sensor. This is sometimes referred to as digital spoofing. Alternatively, a false data stream might be interjected between the sensor and the processing system (i.e., feature extractor and matching unit). This usually involves tampering with the biometric reader.
3. *Override feature extractor*—In this attack, the feature extractor is compelled to produce a preselected feature set or template chosen by the fraudster in lieu of the actual values generated from the sensor. This attack usually focuses on the firmware of the biometric system. One way for this to occur is through the use of a "Trojan horse" that can be inserted into a biometric system, which would create the preselected template. This type of attack could simply disable the system or create a "denial of service."*
4. *Feature representation tampering*—Extracted features can be replaced by a set of fraudulent (e.g., synthetic) features. Between the feature extraction and the matching module, the correct feature set or template is intercepted and replaced by a fraudulent

* A denial of service attack is an attempt to make a normally accessible computer resource or service unavailable to its intended users.

one. This can be accomplished by injecting a false data stream that can iteratively generate new information, retaining only the changes that improve the score until an acceptable match level is achieved. This is sometimes referred to as "hill climbing."* This technique requires access to the system's match scores.

5. *Override the matching unit*—The matching unit is given preselected match scores, regardless of the feature set, which means that a matching unit could be overridden or replaced by a Trojan horse program. This type of attack is subtle, and authorized users may not notice any aberrations.
6. *Tamper with stored templates*—If fraudsters can gain entry into the biometric template database, they can modify one or more templates to authorize an imposter. This type of attack could include the injection of a false template for future use or the acquisition or modification of a legitimate template. This could also be an entry point for a hill climbing attack.
7. *Attack the channel from the stored template to the matching unit*—Data traveling from the template database to the matching unit can be intercepted and modified. In order for the matching unit to compare templates, it must summon the stored enrollment template through a communications channel.
8. *Override the final match result decision*—A hacker might disable the entire authorization system or alter the final decision for a given transaction. It is a form of bypass attack that overrides all decision data and usually injects a false acceptance in all cases. This often involves physical tampering of the biometric reader.

Thwarting the Attacks

Generally speaking, the vulnerabilities of biometric systems are offset by a variety of remedies. For the most part, the biometric industry has anticipated these threats, and most vendors have contrived and tested solutions to successfully counter them. Indeed, most vendors have built-in solutions that prevent most attacks from succeeding and provide remedies for those that do occur. It is somewhat analogous to the security

* Hill climbing in biometrics is an attack whereby a fraudster gains access to a biometric system and at least one user template, and then iteratively synthesizes an image that produces a score to exceed the system threshold. The attacker uses the synthesized image in lieu of the original template to obtain a false acceptance.

software industry, which works tirelessly to provide software solutions that safeguard computer files. Just as computer viruses sometimes succeed in penetrating computer files, attacks on biometric systems are sometimes successful. However, that success is usually short lived as biometric firms offer new, improved countermeasures. The following catalogs some general solutions to known types of biometric attacks:

Attacking the Sensor (Point #1)—This is the primary attack point for most fraudsters. Antispoofing measures include use of accompanying passwords/PINs, smart cards, supervised enrollment,* and enrolling several samples. These are the most obvious low-cost and high-return antispoofing solutions. In addition, one might employ soft biometrics; liveness detection, including challenge–response solutions; or multibiometric systems.

- *Soft biometrics* can help repel attacks against the sensor (point #1) or against modules (point #8) by providing additional information about the individual such as age, gender, height, and weight, but soft biometrics are unable to verify identity by themselves.
- *Liveness detection* helps thwart spoofing at attack point #1. Many biometric systems use antispoofing mechanisms that are aimed at identifying fake properties of biometric samples such as latex, gelatin, or cadaver fingers. Many biometric systems have added liveness detection to their biometric solutions to counter this spoofing by identifying intrinsic properties of liveness, such as a pulse or temperature. Some liveness detection is behavioral, and other methods are physiological. A fraudster cannot use a photograph of someone to infiltrate a facial recognition system when liveness detection is employed, and it is much harder for a fake finger outfitted with a latent print to fool a finger scanning system. Vein pattern recognition systems will reject a severed finger or hand because the biometric reader can only "read" the vein patterns formed by hemoglobin flowing through a live hand or finger. Not only do many vein pattern recognition systems offer built-in liveness detection, but they often use many additional liveness detectors as well.

* User supervision is highly effective in ensuring uncompromised enrollments, but biometrics strength lies in its self-service automation (except during enrollment).

- It should be noted that fingerprint systems are increasingly using liveness detection features. However, in doing so, the addition of that feature tends to impact the pricing of the fingerprint reader and may create a more bulky reader. In addition, some fingerprint liveness detection solutions that have been introduced into the market have been defeated by using thin gelatins or silicon-based artificial fingers. On the other hand, liveness detection for iris and vein pattern systems have so far demonstrated a stouter resistance to spoofing.
- *Challenge–response* can be physiological or procedural. One technique is to provide a physical stimulus and to assess an involuntary reaction. Generally, sensors have enough programmed "intelligence" to support such a challenge and to assess the response. An advantage of systems that use multiples of the same biometric (e.g., fingers, hands, and eyes) is that they can readily employ procedural techniques. One such technique is requesting a random subset of biometric measurements such as requiring the submission of the left hand's middle finger or the right hand's index finger. The randomization is itself security strengthening.
- *Multibiometric systems* are highly resistant to spoofing, as fraudsters must simultaneously spoof multiple biometric traits. Chapter 7 deals with the opportunities and challenges posed by multibiometric systems.

Replay Attack (Point #2)—As surprising as it sounds, some biometric systems lack antitampering defenses such as alarms, special screw heads, control switches, or other mechanisms that either prohibit access or notify the device owner that it was compromised. Antitampering solutions are among the simplest and most cost-effective ways to thwart some common replay attacks. Other remedies include challenge–response systems, encryption/stenography, and time stamping.

- *Challenge–response* systems can help overcome replay attacks. In an image-based challenge–response method, a challenge is presented to a sensor, and the computations for the response are based on the content of the image and the challenge string. This could be a cued challenge whereby the biometric system could further challenge the user if his behavior were considered odd, such as a lack of movement; or the challenge could be random, requiring the user to reauthenticate. Another approach includes the use of a smart card to transfer the verification data to the card for secure on-card matching.

- *Encryption/stenography* provides data with essential protection, since any data can be intercepted when it is transmitted. Stenography is a form of encryption. Stenography refers to hiding key information in the data. One can hide specific information (such as a biometric template) inside other data in such a way that others cannot readily discern the key's presence, much less the data content. Stenographic techniques can be used to thwart attacks at points #2 and #7.
- *Time stamping* is a relatively low-cost remedy to thwart replay attacks. The biometric system would simply time stamp the input signal, and then check the input signal to ensure that it is fresh, within a set of reasonable time intervals, depending on the system and the application.

The next three attacks at *Points #3, #4,* and *#5: Overriding Feature Extractor, Feature Representation Tampering,* and *Overriding the Matching Unit*—All tend to be located inside a physical barrier of some type. Countermeasures include locked doors and various antiaccess solutions, as well as keeping watch on internal staff such that every data change should be subjected to a thorough electronic audit trail.

Tampering with Stored Biometrics (Point #6)—This can occur whether the templates are stored locally or remotely. Most database repositories have security at the point of storage. This includes both physical access security to guard against unauthorized trespass and access to the database, and also information security including an array of antivirus solutions.

- *Cancellable biometrics* is one method to help resist attacks at point #6, the template database. Cancellable biometrics intentionally uses a noninvertible transform to repeat the distortion of a biometric signal.* This enables a legitimate substitution of a transformed template version for matching against a similarly transformed vector. The use of cancellable biometrics reduces the possibility of compromise of the reference templates. IBM's cancellable biometric solution employs algorithms to

* Noninvertible transform is a distortion transform that does not enable the original biometric to be recovered even if the transform biometric data and the transform function are known.

purposely distort images, and it records the distortions. The original image is not stored at all. If a fraudster steals the template with the distorted data, that distortion can be deleted and the proper user can resubmit his biometric sample to create a new distorted reference template.

- *Smart card storage of the reference template* also helps buttress attacks at point #6. Smart cards place the reference templates off-line in the user's pocket, safeguarding them from hacker attacks. Additionally, smart cards are tamper resistant. A smart card can authenticate itself and the card reader by creating a mutually authenticated cryptographic challenge between itself and the reader. This can occur before the ID verification process is initiated and access to the data is granted, ensuring cardholder privacy and preventing inappropriate disclosure of data. Such a procedure would help to defeat the skimming of data that might be used for identity theft or other nefarious activities. Further, by challenging a biometric reader, the smart card can ensure that a previously captured template is not being retransmitted in a replay attack.

Attacking the Communications Channel (Point #7)—Stored enrollment templates must be reviewed by the matching unit, which compares them to the sample template. To do that, the stored templates must travel through a communications channel to the matching unit. Depending on the security of the communications channel, the enrollment template can be intercepted and modified. Encryption is the traditional solution; most biometric systems encrypt transmitted data. Additionally, challenge–response systems and stenographic techniques can augment the system's defense.

Overriding the Final Decision (Point #8)—Use soft biometrics to fine-tune the threshold on the matching score, which can help reduce FAR and FRR errors. Ensure that antitampering mechanisms are built into the biometric scanner.

There are numerous ways in which vulnerabilities are introduced into a biometric system, including system design failures, storage management issues, component defects, and implementation errors. As previously discussed, there are a variety of countermeasures and techniques that can be implemented to nullify and to weaken such attacks, but no biometric system is foolproof. Therefore, all biometric systems should have built-in contingencies in the case of system failure due to a successful attack.

It is important to note that examples of successful attacks on biometric systems, no matter how sensational, do not discredit the value of biometrics any more than the potential for a thief to steal one's house or car keys neutralizes the value of a locked home or car. People will continue to lock their possessions to safeguard (somewhat) against the threat of unauthorized entry.

How much security is enough? The techniques used to thwart attacks on biometric systems add costs and transaction delays. That has to be taken into consideration. The prudent course of action for any enterprise would be to make a risk assessment of the potential threats and to take pragmatic measures to eliminate or reduce those threats to acceptable levels.

The biometric industry acknowledges the vulnerability of its systems to spoofing and other potential attacks, but more proactive initiatives could be taken to address these issues, such as biometric manufacturers hiring universities to try to circumvent biometric products and encouraging systems integrators and other key "middle men" to identify the pitfalls so that they can be remedied before a fraudster can use them.

7

Multibiometric Systems

Security systems are not perfect; neither are biometric solutions. The concept of multiple layering continues to gain traction as the biometrics industry is faced with imperfect systems and imperfect solutions. Whether the focus is on a security system in general or on a specific biometric solution, using a combination of proven techniques has the potential to make any system more robust and alleviates many of its limitations.

A multisystem security strategy is basically the use of two or more levels or types of security techniques. So waving an identity badge (something you have) and entering a personal identification number (PIN) or password (something you know) could be considered multisystem security. The concept is to use multiple techniques or technologies in a layered approach. No single system is foolproof; for that matter, multiple systems are not foolproof either. Nevertheless, multiple security systems used together are generally more resistant to fraud since they employ different techniques or technologies and process information through different algorithms. Similarly, using a multibiometric system (something you are) increases security (e.g., by improving accuracy) and broadens support for and acceptance by the user population by offering alternatives.

This chapter explains the limitations of unimodal biometric systems, describes multiple biometric integration strategies, clarifies how multibiometric systems work, and shares some key issues associated with multibiometric systems including architecture, cost, system enrollment, and training.

LIMITATIONS OF UNIMODAL SYSTEMS

Unitary biometric systems recognize individuals based on a single set of evidence from a biometric trait. As discussed previously, there is no such thing as a perfect biometric technique, and all unimodal biometric systems have limitations, such as noise, expected error rates, nonuniversality, and resistance to circumvention.

- *Noise*—Due to defective or poorly maintained equipment, some sensors may present blurred or indistinct images. The accuracy of any biometric system is highly dependent on the quality of its input. Therefore, "noise" in or around the sensor that causes it to present less than perfect images or signals poses a major limitation. Fingers with cuts across the fingerprinting areas or voices altered by bad colds are examples of noise. The issue with noisy biometric data is that it may cause an incorrect match or deny a genuine match. Unitary biometric systems have a single point of failure in regard to its susceptibility to environmental factors and equipment maintenance.
- *Error rates*—Although all biometric modalities continue to improve their accuracy with improved sensors and more robust algorithms, the rate of error associated with many unimodal biometric systems remains unacceptably high for some sensitive, high-security applications. The three biometric modalities most often used by law enforcement for forensics (fingerprinting) and surveillance (facial recognition and speaker recognition) have relatively high error rates, as much as 1% to 3%, depending on the modality and the application.
- *Nonuniversality*—No biometric modality is truly universal as no modality can claim to apply to every individual in every population. No matter the modality, in any reasonably large population there is a subset of individuals that do not possess the required biometric traits for use with automated systems. The lack of universality is the primary reason for failure-to-enroll situations.
- *Resistance to circumvention*—Unfortunately, it remains possible for imposters to circumvent some biometric systems by spoofing them. In general, unitary biometric systems can be spoofed more readily than multibiometric systems. Vein pattern technology has not been successfully spoofed to date. However, given the relentless advance of technology, appropriate resources, and

determination by a well-organized fraudster group, vein pattern recognition (VPR) system spoofing could occur one day.

In most situations, the deployment of biometric systems is a superior solution to its precursor technologies—badge tokens, keys, PINs, and passwords. However, biometric solutions almost always have a need for improved performance or a broadening of their usability. The degree of needed improvement depends on the application and on the population using the biometric system. In some situations, it may be prudent to consider multiple integration strategies.

MULTIPLE INTEGRATION STRATEGIES

Many of the limitations of a single biometric modality can be overcome through the use of multiple integration strategies. Multiple biometric techniques can be integrated into one security system to adhere to more rigid performance criteria. When used together these multiple techniques or systems can heighten the selection accuracy, can better achieve stringent performance requirements imposed by some high-security applications, can more readily identify imposters, and can decrease the ability of the imposter to spoof the system.

The concept of multibiometric systems is not a new one. It has been used commercially since the late 1990s, with varying levels of success. A multibiometric system is one that uses multiple biometric sources to improve its overall performance. As illustrated in Figure 7.1, it might use any of the following six scenarios (starting from the top and moving clockwise):

1. Multiple sensors producing multiple samples of a single biometric trait to assuage noisy sensor data so that different sensors (e.g., optical and chip-based sensors for fingerprinting) for the same biometric identifier might be used to improve performance.
2. Multiple representations from the same biometric trait (e.g., minutiae-based versus filter-based fingerprint systems) involving different approaches to feature extraction and matching.
3. Multiple samples of the same biometric trait using the same instance (e.g., same finger or same hand).
4. Multiple identifiers, usually referred to as multimodal biometrics (e.g., using finger vein and fingerprint together, or hand geometry and palm vein patterns).

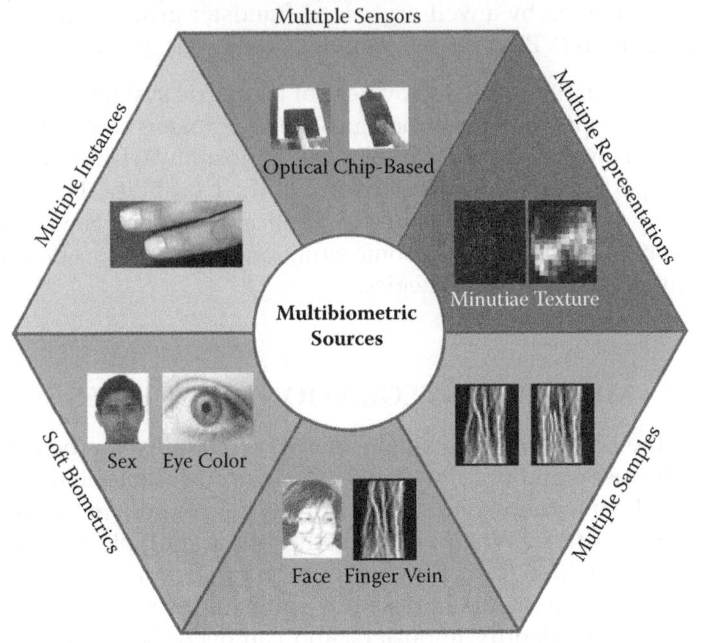

FIGURE 7.1 Multibiometric sources.

5. Using "soft biometrics" such as gender, height, or eye color as a way to strengthen performance and accelerate processing time.
6. Sampling multiple instances within the same modality (also called intramodal) such as different fingers or both palms for vein pattern recognition, which can help avoid spoofing in a challenge–response authentication.

The choice of multiple integration strategies depends primarily on an enterprise's requirements as well as the type of applications supported, the correlations among the biometric identifiers, and, of course, the costs incurred. These techniques provide multiple corridors of security checks that can be performed simultaneously or sequentially. Multiple biometric techniques combine multiple factors of evidence to enable better decisions. By combining the evidence obtained from different sources, biometric systems can overcome some of the limitations of unimodal biometrics and generally improve recognition performance.

When the topic of multibiometric systems surfaces, most people think of multimodal biometric systems that use more than one physiological or behavioral trait for enrollment and identity verification (1:1 matching) or identification (1:n matching). Multimodal systems are arguably the most powerful type of multibiometric system, and they hold great promise for significant performance improvements over unimodal systems. Indeed, multimodal systems analyze the evidence from multiple sources for verifying an individual's identity or for identifying an individual from a database. Generally, multimodal systems provide superior recognition performance over unimodal biometric systems. Multimodal systems offer an extensive set of advantages:

- Reduce the number of false acceptances and false rejections, and thus significantly improve the matching accuracy and the overall performance of the biometric system, often providing a substantial reduction in the error rate.
- Better thwart attempts to spoof a biometric system, as it is difficult to spoof multiple traits simultaneously.
- Extend the range of acceptable environmental conditions (e.g., noise reduction) with which authentication or identification can occur.
- Provide a secondary means for enrollment, verification, and identification, increasing the availability of the biometric system, broadening its population coverage, and minimizing the effects of intra- and interclass variation.

Since no one biometric method can address every member of a large user population, the use of multiple biometric modalities offers alternatives for some individuals (e.g., an individual who has lost his sight cannot use an iris scanner). Using multiple biometric methods can broaden the ability of the biometric system to verify identity by relaxing requirements in one category in favor of another. Thus, if a user has difficulty with one biometric modality, he or she might be able to use the other biometric identifier more easily and with much better results. For some organizations, multimodal biometrics might be used simply to enable a higher degree of inclusiveness for its user community. If different authentication methods are supported by the same system, then a user might choose which biometric method is best suited for him or her, enabling a user to favor one biometric technique over another. Thus, a sightless person might opt for palm vein pattern recognition while an upper bilateral amputee might choose iris recognition. This aspect of multimodal biometrics is extremely important when supporting applications used by the public at large.

As every biometric technique has some key advantages and some specific limitations, multimodal biometric-based systems can take advantage of the performance characteristics of each biometric technique used while overcoming some of the limitations posed by any single biometric technique. The templates for each separate biometric technique could be maintained on the individual's smart card, on a central data repository, or distributed set of servers or scanners.

A multimodal biometrics system could be used in a forensic effort to determine the identity of an unidentified body, in an extensive high-security system to verify one's claim to an identity, or in an enterprise environment to offer alternatives to that portion of the user population that cannot use the primary modality selected. Verification systems tend to use multimodal systems to increase accuracy or improve user convenience and usability. Identification systems tend to use multimodal systems when they can to improve system speed and accuracy. Overall, the use of multimodal identifiers generally improves system accuracy, usability, biometric availability, and user acceptability, and it can decrease system vulnerability.

HOW MULTIBIOMETRIC SYSTEMS WORK

All biometric systems use scores (probability weightings) to express the similarity between a sample biometric pattern and a reference template. The higher the score, the more probable it is a match. Access is granted only if the probability of a match exceeds a preestablished threshold level. Theoretically, imposters generate lower scores than the threshold, and authorized users generate higher ones. In reality, the reverse sometimes occurs, albeit infrequently. Third-party benchmarks statistically measure how often biometric systems grant access to an imposter and how often an authorized person is denied access. A combination of uncorrelated modalities (e.g., the use of vein patterns with fingerprints and iris recognition) generally provides a much greater improvement in false acceptance rate (FAR) and false rejection rate (FRR) performance over a combination of correlated modalities (for example, multiple fingerprint sensors).

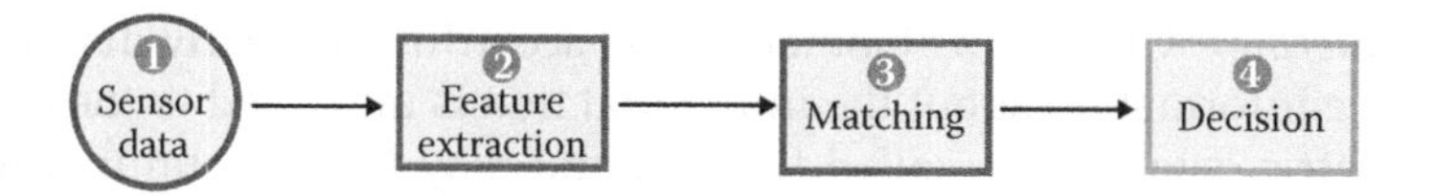

FIGURE 7.2 The four key biometric processes.

In most biometric systems, there are four key processes, as illustrated in Figure 7.2: (1) capturing the biometric trait to be measured in the form of raw data; (2) processing the data extracted into a compressed representation of the trait; (3) comparing the extracted feature set with the reference data, generating a matching score; and (4) using the matching scores to either make an identification decision or to verify a claimed identity. The use of multiple biometric techniques increases the likelihood of a successful match.

Processing Speed Improvement

Chapter 3 discussed how very large databases are dependent on search and retrieval accuracy and how pattern matching for large-scale identification systems can require extensive processing time. For some unimodal systems, a binning technique helps to classify data and thus speed up that process. Binning can also be used with multimodal systems to effect a high level of template classification before performing matching techniques. In addition, two well-accepted techniques accelerate the pattern matching processes for identification applications: pruning and filtering.

- *Pruning*—Narrowing down a large population of potential matches to more manageable subsets can significantly reduce the time required for a pattern-matching process. With pruning techniques, biometric systems generally use the least accurate biometric identifier first and move to the most accurate. Given a potential search database with more than 1 million individuals, a biometric modality, such as speaker recognition, might be used to narrow the search to 150,000 subjects. Then another modality, such as face recognition, might narrow the search to 9,500. Finally, another modality, such as finger vein pattern recognition, would determine a match. This pruning technique requires much less processing time than simultaneously using all three techniques. Of course, this assumes that one has available multiple identifiers on the subjects in question.

- *Filtering*—Classifying biometric data according to clearly discernible information about the individual that is unrelated to the biometric data itself is called filtering and may involve eliminating match candidates by sex, age, eye color, or other distinguishing soft biometric factors that can be stored in an end user's database record. There is a great time savings in not having to review all the records associated with people who do not match the profile provided by the applied soft biometrics. Therefore, if searching for a male over 50 years of age, efficiency suggests that the search eschew all females, and all males under 50.

Binning, pruning, and filtering techniques can be used together if the situation warrants it and if the means exists to do so. All biometric modalities, to include soft biometrics, each can play a role in the identification process.

Accuracy Improvement

There are a variety of strategies to improve the accuracy of multiple biometric systems by combining the biometric data that is collected. Two strategies have been employed with some success: use of a weighted average and fusion of the information garnered from each modality to render a more accurate decision.

- *Weighted average*—A weighted average is straightforward and is considered to be the simplest method for combining scores. This technique averages the various scores from each biometric technique used, weights them by some predetermined formula, and calculates the result. Weighting enables an enterprise to choose how it wishes to emphasize one biometric modality over another. The weighting is effectively an adjustment that accounts for the disparity in accuracy of one modality vis-à-vis other modalities. For example, the error rates associated with palm vein pattern recognition are much lower than those associated with hand geometry. If those two modalities were used together in one system, the palm vein pattern scores should be assigned a greater weight than the scores achieved through hand geometry. Although this technique is subjective, it is quite practical.
- *Fusion*—The use of multiple biometric methods tends to produce a large amount of nonintegrated information. Therefore, a strategy that "fuses together" this wealth of information enables the

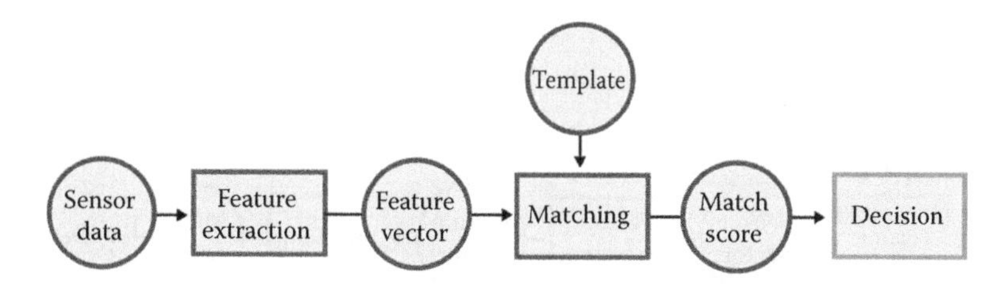

FIGURE 7.3 Authentication process flow.

data from each method to support the other. The information from multiple biometric systems can be combined before or after the data has undergone matching. Fusing the scores of several biometric systems is a well-documented approach to improving overall accuracy. The key to a successful fusion strategy is to select the most appropriate data level for fusion to occur and then to select correspondingly appropriate fusion algorithms.

Industry practitioners tend to classify fusion strategies in accordance with the point in the authentication process at which the information from the different sensors is combined (see Figure 7.3). There are four potential levels at which fusion can occur: (1) the sensor image, (2) feature extraction, (3) matching score, and (4) decision levels. In practice, only two levels have gained any popularity: matching score and decision.

The fusion of data is a form of information integration. Theoretically, the earlier in the authentication process flow that fusion is applied, the better the opportunity is to improve the fusion results. The rationale for this is that the data in the early processes are information rich. However, sensor level and feature level fusion are quite problematic to implement. It is often quite difficult to perform fusion at the sensor and feature levels.

Figure 7.4 illustrates fusion at the sensor level. The sensor module "reads" the biometric attribute (e.g., a hand or finger), and then compiles raw data that is sent to the feature extraction module. In sensor level fusion, the raw data from the sensors are combined. As an example, this consolidation of data could occur if there are multiple reads of the same biometric trait from multiple compatible sensors (e.g., multiple back of the hand vein pattern images of the same hand). With sensor level fusion, the data obtained must be compatible. For instance, reads from multiple sensors of differing quality and manufacture may not be compatible. Theoretically,*

* I refer to the theory of this process because in my research I have not discovered any instances where organizations are in fact fusing data at the sensor data level.

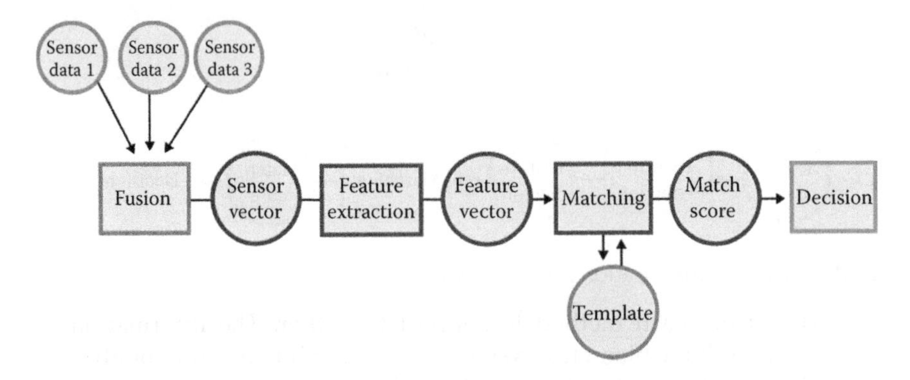

FIGURE 7.4 Fusion at sensor level.

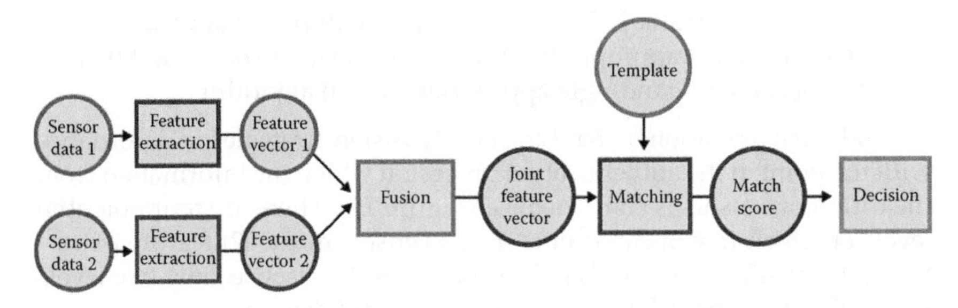

FIGURE 7.5 Fusion at feature extraction level.

data obtained from different sensors would be combined into a joint sensor vector prior to being sent out to the feature extraction module.

Feature extraction fusion refers to combining different target features to produce a new feature set. Figure 7.5 illustrates the process of feature extraction fusion. In feature extraction fusion, feature vectors are combined. An example of fusion at the feature extraction level might occur with features extracted with multiple sensors. When feature vectors are homogeneous, such as multiple finger images, a weighted average of the individual features can calculate a single feature vector. However, this is rarely the case with true multimodal systems. Feature extraction fusion may not be practical or feasible in many situations, and most attempts to fuse multiple modalities at this level have met only limited success. Additionally, most vendors do not wish to release the feature values computed by their systems, rendering feature level fusion problematic. Although it is intuitively appealing to integrate information prior to matching the biometric data, it is difficult to achieve such integration in practice.

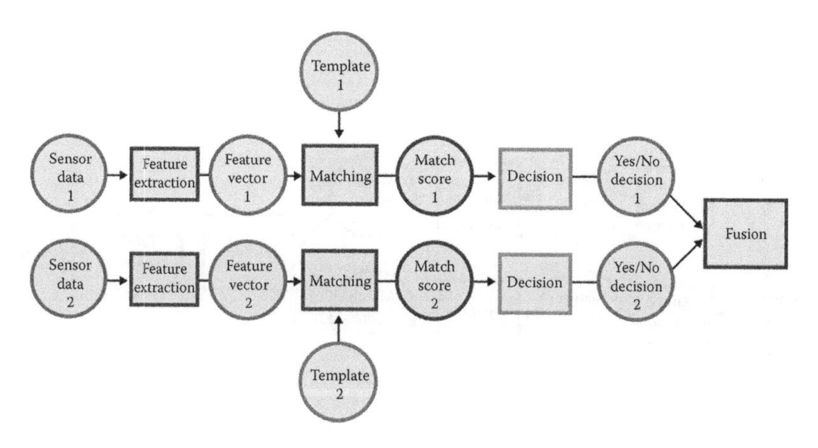

FIGURE 7.6 Fusion at decision level.

Let us briefly skip ahead to decision level fusion, which is depicted in Figure 7.6. At the decision level, each biometric modality renders a separate authentication decision, and then those decisions are integrated using techniques similar to majority-voting schemes. The fusion effort at this level is commensurate to two separate verification processes joined together at their yes/no decision levels. Fusion at the decision level is commonly used but is seen as rigid and somewhat simplistic due to the limited content information available. Indeed, it has acquired some popularity as "layered biometrics." Nevertheless, the limited value of decision-level fusion may not merit the added overhead that an organization would incur in implementing it. However, for certain implementations its use could be quite feasible.

Although there are many fusion scenarios, most multimodal biometric systems integrate data at the matching score level because it offers a strong compromise between the ease in combining the data and better information content, and because it is a relatively straightforward way to combine the scores generated by different matchers. Therefore, matching score fusion is generally the preferred approach for integrating data. Figure 7.7 illustrates this process.

In matching score fusion, scores produced by each modality are combined by a variety of techniques to produce a new score for comparison to the threshold. There are two key approaches in use today for consolidating matching scores: classification and combination. In the classification approach, one can construct a feature vector with individual matching scores, and it is then classified into accept or reject classes. A classification

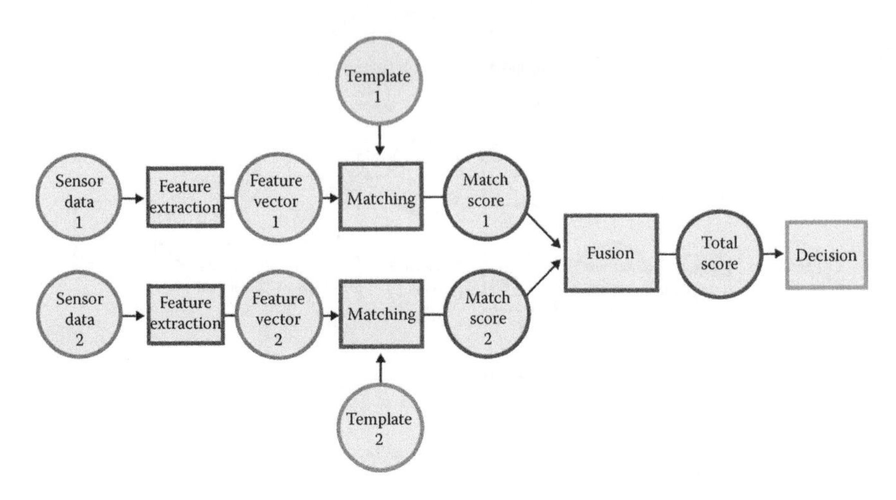

FIGURE 7.7 Fusion at matching score level.

approach might use a decision tree,[*] SVM,[†] or LDA[‡] algorithm to classify the vector as imposter or genuine. With the combination approach, one combines individual matching scores to generate a single scalar score to render the final decision. The combination approach for consolidating matching scores has compiled a superior performance record versus the other levels. Some of the most popular fusion algorithms are:

- *Simple sum rule*—This is basically a weighted average of the raw scores. Matcher scores are summed without benefit of normalization routines. It simplistically assumes that the raw scores supplied by the biometric methods used have a comparable scale, distribution, and strength; there is no rescaling or reweighting to account for matcher accuracy variability. It can be used when there is a high level of noise resulting in some ambiguities in classification. The sum rule assigns the input pattern to class c such that

[*] A decision tree is a graphical representation that displays options, risks, and the decision-making sequence representing a predictive or deterministic model. A decision tree strategy employs a sequence of threshold comparisons to render its verification decision.

[†] SVM is an acronym for a support vector machine, a set of related supervised learning methods used for classification; it belongs to a family of linear classifiers. SVM is based on the concept of decision planes that define decision boundaries.

[‡] LDA, or linear discriminate analysis, models the differences among classes of data. It discriminates between two or more solutions for predictive purposes.

$$f(C_r) = \sum_{i=1} P(C_r / X_i).$$

- *Product rule*—In general, this method is based on the assumption that the biometric traits are statistically independent. Therefore, the input pattern is assigned to class c such that

$$f(C_r) = \prod_{i=1}^{r} P(C_r / X_i).$$

- *Max/min/median rule*—The max rule estimates the mean of the posteriori probabilities by the maximum value, the min rule sets the minimum value of posteriori probabilities, and the median rule estimates the median value by calculating the mean of the posteriori probabilities, as follows:

Max: $$f(C_r) = \max_i P(C_r / X_i),$$

Min: $$f(C_r) = \min_i P(C_r / X_i),$$

Median: $$f(C_r) = \operatorname{med}_i P(C_r / X_i).$$

- *User weighting*—This method computes the combined matching score as a weighted sum of the matching scores. The motivation behind the idea of user-specific weights for computing the weighted sum of scores is that some biometric traits cannot be reliably obtained from some people (e.g., individuals with faint fingerprints). Assigning a lower weight to the fingerprint score and a higher weight to other modalities reduces the probability of a false rejection. The user weighting methods vary and assigned weights may differ among various users.* An example of a user weighting formula is

* There are several versions of the user-weighting calculation. In this case, I depict a formula proposed by R. Snelick, U. Uluday, A. Mink, M. Indovina, and A. Jain in "Large Scale Evaluation of Multimodal Biometric Authentication Using State-of-the-Art Systems," IEEE Transactions on Pattern Analysis and Machine Intelligence, Vol. 27, No. 3, March 2005, pp. 450–455.

$$fi = \sum_{m=i}^{m} w_i^m \; n_i^m \; \forall i$$

where w^m is the weight of matcher m for user i.

Since the matching output from the different modalities is highly heterogeneous, the matching scores generated by the various modalities may not be readily comparable. For example, each biometric system employed might result in different distribution curves or different ranges. In these cases, before combining the matching scores, a normalization step is generally necessary in order to make the results more meaningful.

Normalization

Normalization brings dissimilarly scaled matching scores into a common alignment or standard scale based on the concept of a normal probability distribution. It is a mathematic technique of data transformation and is implemented after producing multiple matching scores but prior to combining them. The process generally results in a more accurate match. Thus, prior to combining different biometric modalities in fusion techniques (e.g., iris scan with finger vein) normalization mutually aligns the distribution curves and achieves better biometric scoring.

- *Min–max* (MM) performs a linear transformation of the original data. This is one of the simplest normalization techniques; it is most useful when the limits of the scores produced are known. It is generally efficient and provides adequate performance, but it may not yield completely accurate results if the data used contains outliers. MM maps raw scores to the [0,1] range, and given s_k matching scores, $k = 1, 2, \ldots n$, such that max (s) and min (s) designate the end points of the score range.

$$n = \frac{s - mean(s_k)}{\max(s_k) - \min(s_k)}$$

- *Z-score* (ZS) is one of the more commonly used normalization techniques. It uses an arithmetic mean and standard deviation to normalize data; therefore, a priori knowledge regarding the average score and score variances of the matcher is needed. It is considered generally efficient and tends to work exceptionally well if the scores of each modality used follow a Gaussian distribution,[*] but this technique may not achieve similar accuracy if the data used contains outliers since the mean and standard deviation are sensitive to outliers. ZS normalization transforms the scores to a normal distribution with an arithmetic mean (μ) of 0 and a standard deviation (σ) of 1:

$$n = \frac{s - mean(s)}{Std(s)} \text{ or } n = \frac{s - \mu}{\sigma}.$$

- *Decimal scaling* (DS) is best used when scores of various matches differ logarithmically. In other words, the scales markedly differ as when one biometric matching unit supports a range of 0 to 1 and the other supports a range of 1 to 100:

$$s' = {s}/{10^n}, \ n = \log_{10} \max\{S_k\}.$$

- *Hyperbolic tangent* (Tanh) is generally efficient and provides adequate performance. It is very robust in handling outliers; however, it has been demonstrated that to work efficiently the parameters must be selected carefully. Tanh maps the raw scores to the (0,1) range, where μ (s) and σ (s) are the mean and standard deviation estimation of the score distribution, respectively:

$$N = {1}/{2}\left[\tanh\left(0.01\frac{\left(s - \mu(S)\right)}{\sigma(S)}\right) + 1\right]$$

- *Median absolute deviation* (MAD) is insensitive to outliers in a distribution, which renders it a very robust solution. However, MAD does have a lower efficiency when compared to the mean and standard distribution estimators. If the scores of each modality used do not follow a Gaussian distribution, then MAD may be a poor estimator.

---

[*] Gaussian distribution is a bell-shaped curve corresponding to a population that has a normal distribution.

$$N = \frac{s - median}{MAD}$$

$$MAD = median\left(\left|\{S_k\} - median\right|\right)$$

If implemented correctly, matching score fusion can improve accuracy, better thwart fraudsters, and increase usability. Implemented incorrectly, a multibiometric system might actually experience performance degradation in comparison to a unimodal solution. Further, multimodal systems potentially have a higher cost of ownership, can increase user inconvenience, can decrease user acceptance, and can exacerbate privacy issues.

KEY ISSUES

Some key issues with multibiometric systems are: (1) architecture, (2) the total cost of ownership, and (3) user enrollment and training. Let us examine each briefly.

Multimodal Architecture

Multimodal architecture usually refers to the sequence in which multiple biometric reads occur and are processed. System complexity can be a major drawback to multimodal systems. From a user perspective, multimodal biometrics can be categorized as synchronous and asynchronous.

Synchronous systems, also referred to as parallel schemes, enable the users to interact with two or more biometric techniques almost simultaneously, for example, palm vein recognition and hand geometry, or speaker and face recognition. In a synchronous design, as illustrated in Figure 7.8, the modalities used operate independently of one another, even if they are combined in one device. There are some practical reasons why it can be difficult to use synchronous multimodal biometric systems including coordination among the multiple sensor systems as well as some aspects of training. Parallel architectures, such as the one depicted in Figure 7.9, generally offer higher accuracy levels because they obtain more data about the user in order to render a decision. The improved accuracy levels of this architecture may result in lower error rates than those garnered by the same modalities deployed in a serial scheme.

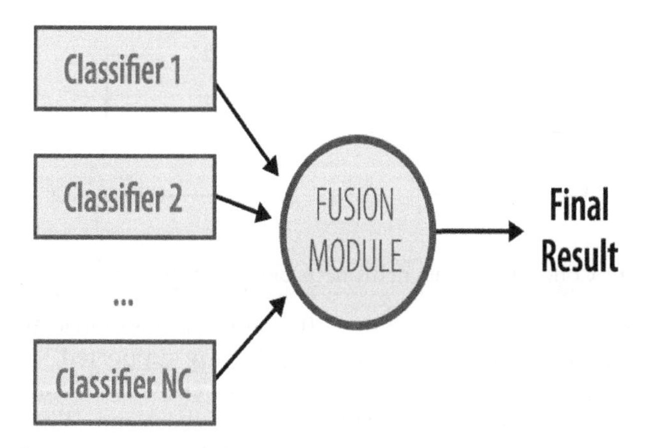

FIGURE 7.8 Concept of a synchronous biometric system.

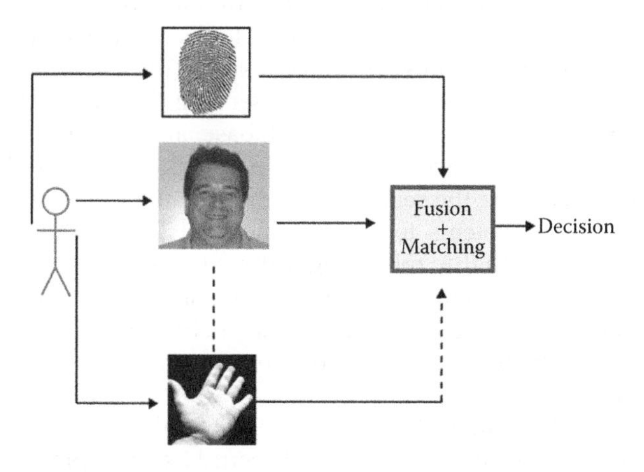

FIGURE 7.9 Parallel biometric architecture.

Asynchronous systems, also referred to as serial schemes, enable the user to interact with different biometric techniques sequentially, and the outcome of one modality can affect the processing of subsequently encountered modalities. So a user might use hand geometry to enter a facility, then use a fingerprint biometric to enter a more restricted floor and then use iris scanning to enter a top secret room within the facility. Or all these systems can be combined from the beginning to tighten overall security. Such biometric architectures are more commonly used

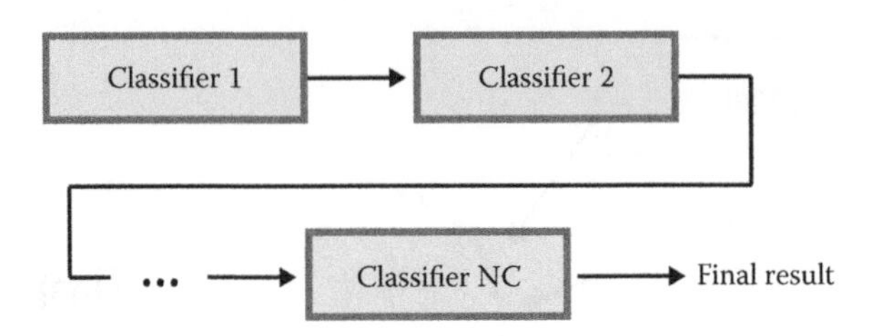

FIGURE 7.10 Concept of an asynchronous biometric system.

in highly secure facilities. Of course, the choice of biometric architecture is highly dependent on the application(s) that are supported.

Asynchronous systems can improve user convenience and support faster searches for large-scale identification applications. When an asynchronous system verifies an individual's identity with high confidence at the first modality, the user might not be required to provide biometric samples to the successive modalities in the architecture. Moreover, such a system could be designed to enable the user to decide which modalities he or she would use. This could significantly expand the usability and acceptance of a highly secure biometric system serving the public. With regard to an identification task using a large database, an asynchronous system can use the successive outcomes of each biometric encountered to prune the database, rendering the search more efficient and potentially faster.

Asynchronous systems tend to be fast and highly convenient, but they usually require highly robust algorithms to handle each sequence of events. Such systems would be preferred in most public-facing situations. Figure 7.10 depicts the serial nature of an asynchronous biometric system and Figure 7.11 illustrates the decision process provided by this biometric architecture.

It is possible to combine both parallel and serial architectures into one system. However, given that heightened security and improved accuracy may be the key reasons for developing a multimodal system in the first place, most systems in place today tend to use the parallel architecture.

Total Cost of Ownership

The core reason explaining why multimodal biometric systems have been limited to highly secure facilities is that they tend to be expensive. An enterprise that employs multiple biometric identifiers has accepted, purposely or unintentionally, the increased computational overhead,

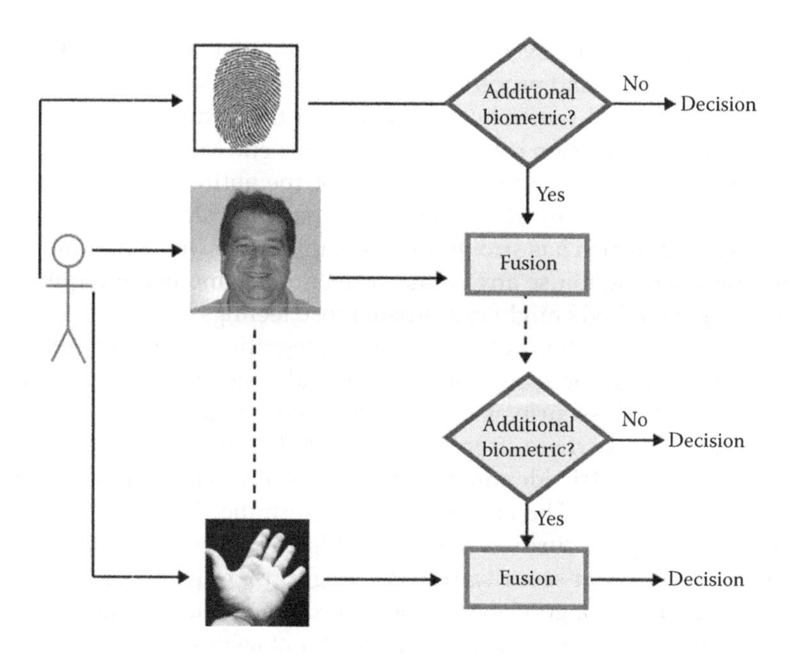

FIGURE 7.11 Serial biometric architecture.

integration costs, greater sensor acquisition costs, attenuated project management costs, and ongoing maintenance and other information technology (IT) expenses. Increased system processing impacts the overhead costs as well. Moreover, the enrollment time for multimodal systems is usually attenuated and that can increase the cost of staff training and enrollment. The cost of the total system is often greater than the sum of the parts. Therefore, a key consideration for organizations contemplating biometrics is whether the benefit derived from implementing a multimodal biometric system exceeds the upfront investment and the ongoing operational and maintenance costs as measured in both hard and soft dollars.

User Enrollment and Training

Multimodal systems generally improve coverage of a given population by reducing the failure to enroll (FTE) rate. Admittedly, multimodal systems usually require more time and effort for users to enroll and to verify themselves. Multimodal systems also introduce a new level of complexity. As has been alluded to, it is generally more challenging to train users to use multiple biometric systems. Each biometric system tends to have its own idiosyncrasies and procedures, which can be quite overwhelming for users to master and remember, regardless of their education, unless the training and supervision is thorough and excellent. Moreover, FTE in one modality could translate to a limitation for the entire biometric system. Additionally, the user must reveal multiple personal identifiers in a multimodal system, which has strong privacy implications. Indeed, some users may resent having to use any biometric technique and being required to use multiple methods might exacerbate those feelings.

Due to considerations of information integration, total cost, and end-user training, past successes with multimodal systems have been limited primarily to high-security facilities where cost issues are not a prime consideration and where users are motivated to master these systems. Multimodal biometric identifiers can successfully retain high-threshold recognition settings. The enterprise can determine the level of security required. For high-security scenarios, multiple biometric identifiers might all be used; for lower security, the user might use only one of the identifiers. In either case, a good system design coupled with logical selections of biometric modalities would be highly advantageous.

COMBINING BIOMETRIC METHODOLOGIES

Multimodal biometrics suggests a range of differing modalities used in combinations that may not be initially intuitive. The combinations selected depend on several factors, including the applications supported, the accuracy needed, the fusion level achieved, and the techniques selected. Some basic combinations include:

- Facial and speaker recognition
- Facial and iris recognition
- Hand vein pattern and hand geometry recognition
- Vein pattern and facial recognition
- Vein pattern and iris recognition
- Fingerprint and finger vein recognition

This list is not meant to be comprehensive but simply suggests possible modality combinations. The speaker recognition and facial recognition combination is intuitive. Other modality combinations, although not initially obvious, might together provide significant performance improvement that could not have been achieved singularly. Fingerprint and finger vein pattern recognition provide an interesting combination.

Fingerprint and finger vein recognition biometric methods are totally different technologies, and yet they share much in common and could potentially work well together. Not only do they both focus on one's finger, they can be used with any finger or a combination of fingers. For example, to tighten security, an enterprise might require the use of one finger from each hand, or it might require the index finger and the middle finger from either hand. Both technologies can support this.

Since each of these identifiers "reads" a separate part of the finger, there is no interference with one another. Although humans have unique prints on all parts of their fingers and palms, fingerprint scanning systems only focus on the top one-third of a finger. The "tip" of the finger is that portion of the finger above the upper distal interphalangeal joint. As illustrated in Figure 7.12, each human finger has three parts: (1) a distal phalanx (tip of finger), (2) a middle phalanx, and (3) a proximal phalanx (where the finger joins the palm, the metacarpals). Finger vein pattern recognition systems read the vein patterns that form in a finger's middle phalanx. Although finger vein pattern recognition technology could read any section of one's finger, the middle phalanx is the most convenient. In this fingerprint–finger vein bimodal system, the reads can occur simultaneously and in one swift action in laying one's finger on the reader. To the user, he or she is just getting a single read. Because the use of this biometric system would be so intuitive and easy to use, it obviates the need for substantial initial or remedial training.

Because the readers of both technologies are so miniature, their joint use continues to enable multiple form factors as well as provide a small "footprint" wherever they are used. Combining the multimodal fingerprint and finger vein sensors in one reader and creating one single system significantly reduces costs. The information integration can readily occur at the matching score or decision levels, or potentially even sooner. Last, this multimodal system can either heighten security by requiring a positive match for both biometrics, or it can broaden acceptance by enabling its uses to support one system or the other. Thus, people with poor or faded fingerprints can use finger vein, and those with vein anomalies can use the fingerprint solution, or the acceptance threshold can be set to support both systems to heightened security.

Although the finger vein pattern recognition biometric is generally a 1:1 identity verification method when in stand-alone mode, it can support 1:n opportunities when it is part of a query database supporting

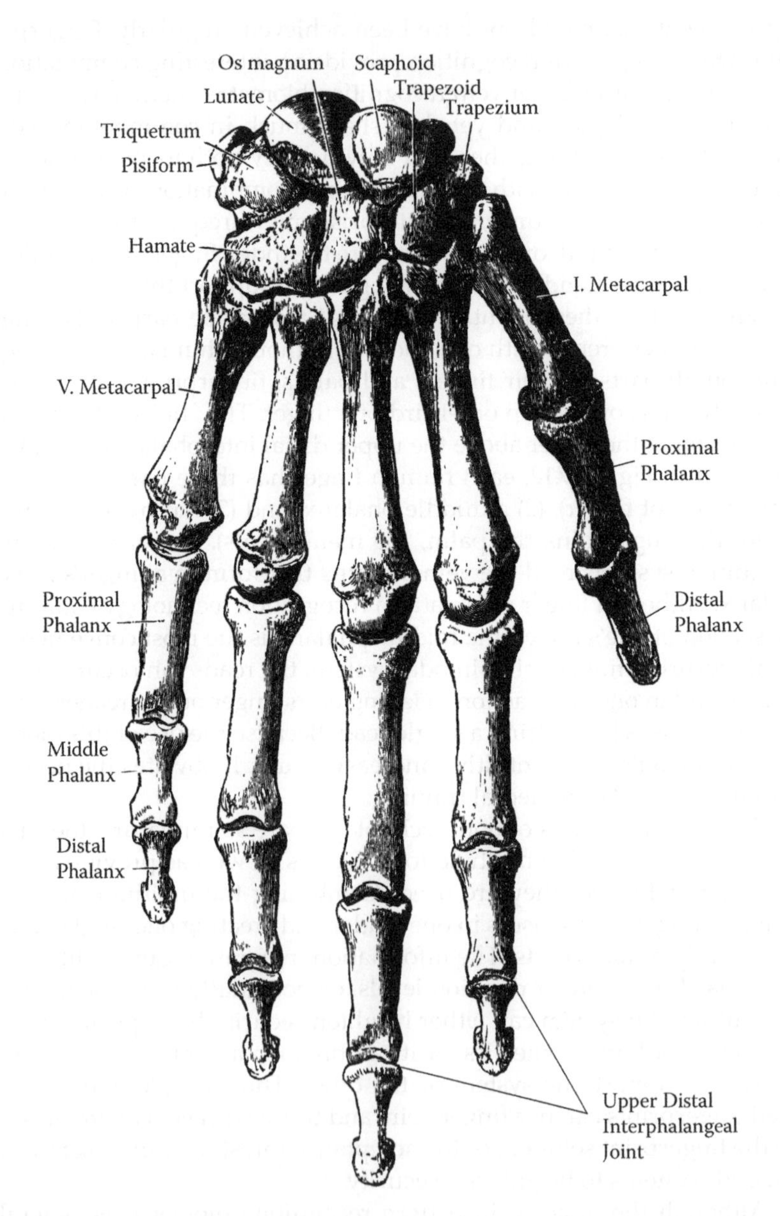

FIGURE 7.12 Anatomy of the human hand.

multimodal biometrics. Using finger vein pattern recognition in this way will extend the utility of the biometric to markets that otherwise would be closed to it.

A combined fingerprint and finger vein system could maintain its templates either on a smart card or other individual token devices, which could be inserted into (contact) or waved at (contactless) a smart card reader for matching to the live template. Or the templates could be maintained in a data repository at a central location or distributed locations.

The multimodal device could work together in a fashion similar to the way each device works separately today. The user would place his finger on the reader using the built-in guideposts. The reader would have a glass-like surface to aid in imaging the tip of the finger, while the sensor corresponding to the finger's middle phalanx (middle part of the finger) would be contactless with the charge-coupled device (CCD) camera positioned several millimeters from the finger. Both imaging sensors would filter the images, produce digitized images for comparison to the stored templates, perform the matching sequence, and provide a response, depending on the acceptance thresholds and requirements of each system. Figure 7.13 depicts one possible integration of fingerprint and finger vein, each with separate matching algorithms yet allowed to combine for a "decision" to authorize access or not.

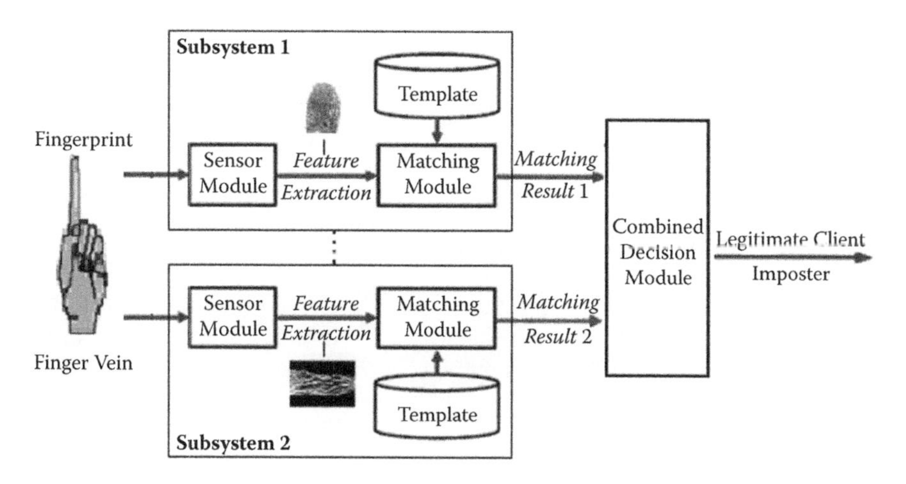

FIGURE 7.13 Integration of multiple biometric technologies. (Courtesy of Hitachi Ltd.)

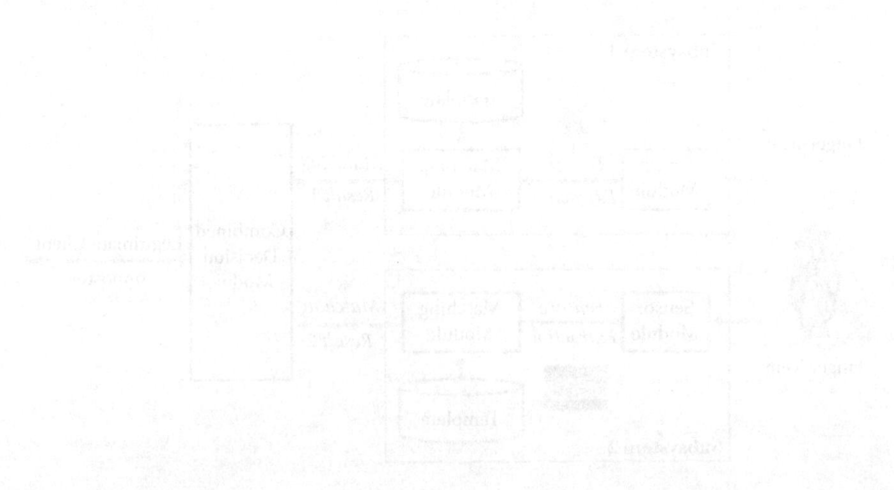

8

Plan Your Biometric System

A How-To Guide

A successful biometric system begins with a well thought-out plan and is nurtured by solid project management. A biometric system project plan is essentially similar to other project plans, and any good project manager can readily manage a biometric system project. The designated project manager just needs to fully use the skill set that made him or her successful in other areas. The project management principles are the same.

Let us begin with overall project feasibility, sometimes referred to as an opportunity assessment. The best way to start a feasibility review is by determining the primary business needs. Write out what you are trying to accomplish with the biometric system. That will help clarify the situation, and it will enable you to share your thoughts with others and obtain confirmation from project participants and other stakeholders. Also note what types of applications you want to implement (e.g., physical access, logical access, accountability, membership, or time and attendance). If there is more than one, prioritize them. Focus on a few core issues that your new biometric system can solve. Prioritize other opportunities for future efforts. Next, define the stakeholders for this project. That is, list everyone, within and outside the enterprise, who will be affected by the project. This may have geographic implications as well.

Your feasibility assessment should list project constraints. Determine if you have any technical or operational constraints. Are you replacing an existing security system or adding a biometric dimension to an existing security system? Either way, what kind of information do you want your new system to provide? To what degree will your new biometric system require design, development, and testing? Will it interface to your legacy systems? How do you plan to train your employees? How will you accommodate guests?

Next, consider financial feasibility. Like most projects, biometric solutions require initial investments of funds as well as funding for ongoing support activities. Basically, will you be able to properly fund the project from completion through ongoing support? Has a cost–benefit analysis been performed? This cost–benefit analysis should be in alignment with the scope of the biometric project and the standards of your company. Project benefits are often intangibles such as efficiency improvement, heightened security, better information flow, and improved safety. However, it is a good practice to quantify them whenever possible. Your planning costs should include new hardware, application software, systems support, engineering support, training, and ongoing maintenance.

You should know how the biometric project fits with the short- and long-term strategic plans for the company. Biometric systems can strongly impact building management, information technology (IT), human resources, security, and legal, to name some potentially affected departments. Does this project conflict with or complement other company initiatives that are being planned or are already underway? What processes need to be changed as a result of the biometric project?

Once the project feasibility appears to be on track and the cost–benefit analysis appears positive, then the next step is to prepare a project recommendation and name a project sponsor. The project recommendation is often a brief presentation that is communicated verbally or in writing to senior management. Typically, it is a summary of the feasibility issues, the cost–benefit analysis, and your recommendation to proceed with a biometric solution. Depending on the organization, it might include a risk assessment section as well. The purpose of the project recommendation is to demonstrate how this biometric system project aligns with the company's strategic plan, as well as to convince senior management of the company to support the project. If accepted, the senior management should formally appoint the project sponsor. This is the key senior person within the organization who will sponsor the project and be responsible for its funding. If the project is being undertaken by a contractor, then the sponsor is usually the key contact person in the client organization. If the project is being undertaken by the client organization with its own resources, then the sponsor is often the executive to whom the project manager reports or it is the executive most interested in and most affected by the project implementation. What is critical is that the biometric system project be sponsored by an executive who has some financial responsibility for it.

PLAN THE PLAN

Once the biometric system project is authorized and any required sign-offs are obtained, it is time to develop the actual plan. The size of the plan will depend on the scope of the project as well as the organization's standards for such documents. This plan should be committed to writing, even if the biometric system project is relatively small. A two- to three-page plan is preferable to no written plan at all because things can change and begin to get complicated even in what seems like a simple tracking application. Moreover, most plans quickly become too detail oriented for someone to try to remember all the specifics. Another important benefit of a written plan is the ability to share it with others. It helps all the project stakeholders to better understand what is planned and enables them to make advance adjustments and plans of their own. There are nine basic components needed to start the project plan as shown in Table 8.1. Each is discussed in the following sections.

Project Goal

The project goal is the first item in the project plan. This goal could also be considered a project mission. It is essentially a high-level statement

Table 8.1 Biometric Project Plan

Number	Item	Action
1	Project Goal	Define it
2	Project Objectives	State them
3	Project Scope	Determine it
4	Project Deliverables	Define them
5	Project Constraints	Evaluate them
6	Project Approach	Define it
7	Required Human Resources	Determine what they are
8	Project Assumptions	List them
9	Critical Success Factors	Develop them

regarding why the project is being undertaken. The stated goal will serve to help guide the project team in completing the rest of the plan; and it is a key consideration at the end of the project during the evaluation phase to understand the original intent and purpose of the project. The goal statement can be as broad or as specific as your project scope dictates. What is important is that the goal statement captures and reflects the project's professed end result.

Project Objectives

Project objectives are high-level statements. They differ from the project goal in that objectives answer what, not why. Most biometric system projects can be covered in four to seven project objectives. To be considered well-grounded objectives, each should be specific, attainable, measurable, and time oriented. Sample objectives might be expressed as, "Establish new access control procedures by Oct 20xx," or "Complete the enrollment of the staff one week before implementation."

Your project objectives should cover each key aspect of the project including procurement, design, and implementation, as well as support functions. Thus, if you list the objectives one after the other, they should form a sort of bulleted synopsis of the entire project. A key point is to not confuse project objectives with success criteria. For example, the hackneyed phrase "complete the project on time and under budget" is not a project objective; it is a criterion for success. Finally, for many organizations each project objective becomes a milestone that can be measured. Tracking milestones with their respectively stated timelines helps an organization to track the progress and status of the project.

The following is a hypothetical set of project objectives to implement a biometric system pilot for a physical access application:

- Select a biometric technique and a vendor based on our needs (e.g., applications, costs, and so forth) and the documented performance of the biometric solution.
- Integrate and test biometric system scanners with legacy infrastructure.
- Perform an operational evaluation of the selected biometric system prior to installation.
- Install the biometric system.
- Perform acceptance testing after installation.

- Enroll all staff participating in the physical access pilot.
- Evaluate the results 3 months after the pilot and enterprise-wide launch.

Project Scope

Many readers may recognize the project scope by its other well-known name, the statement of work (SOW), which is a set of statements that describe the tasks to be accomplished throughout the project. Theoretically, the project scope statements, when summarized, should describe the entire biometric system project from end to end.

A common term that project planners have come to fear is the infamous *scope creep,* referring to tasks that some stakeholders believe were implied in the scope. It often occurs when a client expects more than a vendor is providing. However, it can also occur within organizations, as when project stakeholders get so excited about a project they start adding their wish list of tasks that they believe were implied or which they simply want to add to the project. Therefore, it is very important in this section to list scope exclusions. Although this seems a bit obtuse to some, the reason why the project plan should explicitly state exclusions is to remove ambiguity and clarify the tasks that are and are not within the scope. Then, if a stakeholder feels that an important task has been omitted, the project participants can discuss it and make a conscious decision to include or exclude that task.

The scope statement establishes a common understanding of the project's purpose for team members and stakeholders. It will contain the criteria that are used at the end of a project to determine whether the project was completed successfully. Contents of a scope statement include:

- The biometric system project overview, including a description of the end product of the project.
- The goals of the project.
- A list of the project deliverables.
- A list of the project requirements.
- A list of any exclusions.
- High-level time and cost estimates.
- Roles and responsibilities.
- Assumptions about the project.
- An outline of any constraints that may hinder progress.

Project Deliverables

Project deliverables are the intermediate and end results of the project, the tangible benefits of the project. These deliverables can be categorized and further subcategorized as needed. In the plan they can be easily listed as bulleted items. Many ideas for deliverables come from the project objectives. Just revisit each objective and try to determine what deliverables are needed to accomplish that objective. Thus, a biometric system project manager might list the following as some deliverables:

- Selection of a site for the biometric system pilot.
- Selection of biometric modalities and vendors.
- Completed survey of current processes.
- Published recommendations for new processes.
- Training of each staffing shift that will provide enrollment support.
- Production of biometric system support manual.

Project Constraints

A listing of the project constraints is primarily a management tool. By listing them, you share these project limitations with project stakeholders. It will help manage expectations, and when it is time to perform a project evaluation; and it will serve as a written reminder of the conditions that were faced. Try to list your project constraints in a bulleted format in as detailed a manner as possible. Because every project has resource limitations, it is inadequate to just list that category. To be useful, project constraints need to be specific, measurable, and realistic. Thus, one might list the constraints in terms of available staff, funding limits, time constraints, and even impending events.

Project Approach

The approach is merely one's strategy to accomplish the biometric project goal and objectives. For example, one strategy that is often employed for biometric projects is to divide the project into phases. Phasing a project allows an enterprise to move slowly but deliberately into a new technology while limiting the opportunity for any catastrophic failure. Another project approach might be to outsource the initial implementation for the pilot and then take gradual control of the effort with in-house resources, giving your staff an opportunity to learn the biometric system first and then gradually take over the installation and support requirements. Other strategies might include installing the new biometric system in a new facility with new workers (where there is limited opportunity to fall back on previous access control practices) or to install it in a smaller, self-contained enterprise as an initial biometric system pilot. Whatever strategy is chosen, it should support the organization's project goal and objectives.

Required Human Resources

A list of needed internal resources should be made, in addition to the list of external resources noted earlier. The list should include the resources' skill levels, their project roles, and the time intervals in which they will be needed. An accurate job of listing resources requires a thorough understanding of the biometric system project. Thus, if you list two engineers for the design stage, you should explain the skill levels needed, the roles that they will play (for example, a chief network designer versus a programmer), and the time frames you will need them. If a key person is performing multiple roles at various stages of the project, then you should note that. A thorough documentation of the human resources needed will make the creation of a project management report much easier as well as increase the accuracy of that tool.

You will want to work closely with your human resources department to ensure the establishment and execution of an enterprise privacy policy. Even if your company has a current privacy policy, it may need some tweaking in addressing potential issues regarding the use of biometrics. The key is to publicize the privacy policy within the enterprise, adhere to it, and to ensure that all employees are educated regarding the implications of a biometric system.

Project Assumptions

You should list project assumptions for several reasons. First, it is a communications tool. You will want all the stakeholders and especially your management to fully understand the project. Communicating assumptions is a key step to articulating your concept of the project. Second, a listing of assumptions helps to validate the intent of the project. Finally, separating assumptions from facts helps clarify the project.

Listing assumptions is hard to do. On the one hand, you want to be very detailed, but on the other, it serves no purpose to clutter your plan with lengthy lists of trivial assumptions. The aim is to strike a middle ground of useful, high-impact assumptions. Also, it should be noted that assumptions are somewhat similar to constraints and exclusions. It is okay to have some duplication among those categories.

Critical Success Factors

Success factors are simply the conditions that ensure successful project completion. They include time factors, budget or cost–benefit ratios, employee and customer satisfaction, or simply meeting objectives. The key is stating up front what your success factors are. This helps you gain support from the stakeholders, and it defines how your biometric system project will be evaluated upon completion.

CONTROL THE PLAN

Once you have penned the nine components of your biometric system project plan, you should consider a set of control plans (see Table 8.2) that often accompany the overall project plan. The larger the project, the greater is the need for these control plans. Sometimes, control plans are simply appendices to the overall plan. They can be quite lengthy but often consist of only a few concise pages. A biometric system installation can be brought to successful completion without any of these control plans, but most projects benefit from their creation and use. They include, but are not limited to, a communications plan, a change control plan, a quality management plan, and a procurement plan. They can be vitally important whether an organization is managing its entire project in-house or working with an outside vendor.

Table 8.2 Control Plans

1. Communications
2. Change control
3. Quality management
4. Procurement

Communications Plan

A communications plan defines who communicates to whom about what and when. There are many lines of communications between your staff, your management team, your vendors, and your project stakeholders. Not defining the lines of communications or who gets reports, and not stating level of detail or the frequency of reports can sabotage the success of any project. Even a one-page chart that details the flow of information, the periodicity of the reporting, and the manner in which you do it (face to face, e-mail/letter, or telephone call) is a very valuable communications tool. Reports can be status reports, exception reports, progress reports, schedules, and forecasts—whatever the organization feels that it needs.

Change Control Plan

Many readers will see this and exclaim that their projects do not need a change control plan. They may be right. However, given that its purpose is to minimize scope creep and to document changes, ensuring that all appropriate authorities in the organization agree upon those changes, they might reconsider its importance.

Throughout the life of the project, there will be many suggested changes by an array of well-meaning stakeholders. Some changes will be absolutely necessary, such as when an unforeseen government regulation surfaces that may not have existed when the project began. Other changes will be politically motivated. Still others will seem reasonable at first, but their value may diminish along with your desire to implement the changes as you consider their impact to the overall project.

A biometric system project manager must define what constitutes a change in the project. Is a particular variation in a project deliverable a project change? What about a task variation? Are there any new constraints or product substitutions? What about goals? Some changes could have a huge impact on the project, its completion, and its costs. One reason to support a change control process is to help determine the potential impact of a change and whether that change is necessary or even desirable. Formalized changes usually require mutual agreement from various parties. Another key reason for a change control plan is to standardize the procedures in advance of a suggested change so that various stakeholders are not alienated when they eagerly provide their suggested changes, which the project manager might not eagerly accept. Finally, during the evaluation stage, no matter whether the project was marginally successful or significantly so, the project manager may be asked to account for changes from the original plan. Having a formal change process and documented rationales for the changes helps the project manager respond to queries from his management or other stakeholders.

Once you determine what constitutes a change, it is important to create a change notification process, a brief outline of who notifies whom of a potential change, how that change will be formatted in writing, and how the various stakeholders will be notified about the potential change.

The easiest way to accomplish this is to design a change control form that includes all the pertinent information and enables an intuitive flow of information and approvals. Additionally, the change control process should define the levels of authority that can approve a change. This is often handled in terms of the authority that approves certain funding levels, as it is assumed that many changes have cost implications. Equally important is defining the feedback process to notify the stakeholders of the disposition of the change request. Indeed, each formally suggested change should be assigned a control number by the project manager. In addition to keeping a copy of each adjudicated change, the project manager should maintain a spreadsheet log of all change requests, a brief description, their costs, and the adjudication of each change request.

Once those issues are determined, it is a straightforward process to write a brief change control plan for your biometric system projects, defining the procedures and change request format, the adjudication process, and the dissemination of the results. In this way, suggested changes will have limited impact on the project and on project management.

Quality Management Plan

The quality management plan directs how the project management team will address quality issues. The plan describes the quality processes that affect the project deliverables. Quality standards are defined by each organization. However, there are two key components to a quality management plan: quality assurance and quality control. Generally, quality assurance is focused on preventing quality issues, whereas quality control is focused on correcting quality issues. Very often, organizations will have internally published standards for quality assurance and control processes. If so, it is easy to simply refer to the company standards in your project plan.

Quality assurance processes describe how you will assure quality (however you define it) in your project and in its deliverables. Thus, a quality assurance process generally defines how you will audit the biometric system project and its deliverables, how you will know whether your organization's quality standards are being met, and how you will invoke corrective action if they are not.

Quality control describes how you will inspect the deliverables, in what circumstances, and with what resources. It explains specific corrective actions to be taken if problems are identified.

Once you determine how to assure and control quality and how to evaluate it in terms of deliverables, then it is a not-so-daunting task to create a quality control plan for your project. The document can be very detailed with subsections or it can be a straightforward plan with bulleted statements.

Procurement Plan

The procurement plan is a must when using an outside contractor to implement your biometric system; but it is still a good idea when implementing with in-house resources, since you still have to procure some biometric equipment. This plan describes how you will notify vendors of your interest in their services or products, how you will select vendors, what your selection criteria are, and how you will notify a vendor of his selection. Additionally, it often includes your organization's standards and procedures (or refers to them if they are internally published). Last, the procurement plan enables ongoing documentation of the materials, equipment and services procured, the terms and conditions, and any service level agreements (SLAs) that your organization requires.

EXECUTE THE PLAN

At this point, a lot of hard work has gone into your plan for a biometric solution. Given that you have the approvals from your management team, you are now ready to assemble your project team and start the project.

Kickoff Meeting

Most seasoned project managers like to start with a *kickoff meeting* or meetings (depending on the size of the project and the audiences, and the geographical venues involved). The meeting can be a teleconference or in person. The importance of the meeting is to share information, level-set the project, and manage expectations. The kickoff meeting should have a formal agenda and should include such items as a project overview that includes the project scope, goal and objectives, project roles and responsibilities, schedules, and a discussion of the control plans. Another important aspect of the kickoff meeting is that it provides tangible evidence to all stakeholders that the project is underway.

Monitoring Progress

Once the project is officially underway, the focus of the project manager is to monitor the work being performed and make adjustments. You should implement and follow the published control plans. You will want to communicate your progress to the appropriate stakeholders in a timely fashion, whether the news is good or bad. Bad news does not get better with age, and in the absence of communications with stakeholders, they will assume the worst. Keeping everyone well informed limits rumors and helps you to manage expectations.

Monitor especially your milestone achievements and your costs. Milestones tend to be the salient events or critical path deliverables that have high visibility within the organization. It is likely that you may be late on a few milestones. That happens in the best-run projects. You should determine what caused you to miss the milestone and make adjustments to preclude future repeats. Tracking project costs will help keep you within budget and will provide invaluable benchmarks for future projects.

You will need to update your project plan periodically. The important thing is to understand why you make changes to your plan. There may be implied requirements that were never documented per scope creep or some approved changes might have a greater impact than originally understood. Whatever the issue, you should document it, understand it, and adjust to it.

Reporting Project Status

Producing and distributing timely reports will keep your management and your stakeholders apprised of your progress and will help you immeasurably in the long run.

The standard report for projects is the status report. The periodicity, format, and distribution list will vary by organization. Generally, these status reports regarding your biometric system are sent to the senior management, the project sponsor, or sometimes to the project steering committee or other oversight body. Additionally, most project managers prefer to hold weekly or monthly status meetings.

The audience for the status meetings includes the project team members and often the project champion or other senior management. Some meetings include all stakeholders; or alternatively, the project managers simply hold two meetings: one for their project team (weekly) and one for stakeholders (monthly or quarterly). The purpose of the team status meetings is for members to interactively update one another and problem solve to remedy issues that occasionally surface, such as availability of a given resource or an unanticipated requirement. The purpose of the stakeholder meetings is the communication and the dissemination of information. The value of an in-person meeting to the stakeholders is the opportunity for them to ask questions of the project manager. The value to the project manager is that he can position how information is presented and help control any unfounded rumors. It is a good practice to distribute an updated schedule or Gantt chart at these meetings, as they convey the project status in a concise manner.

DESIGN THE SOLUTION

It is here in the design the solution stage that one has to accurately determine what biometric system components (hardware and software) are needed and how they will be put together. The application has a stated goal or purpose, but the solution will provide the details of how the goal is achieved.

Business Issues

First, review your business processes related to the application for which a biometric solution is needed. The project manager should not do this alone. He should form a team of biometrics-savvy employees to work with him to help establish the project strategies, to evaluate the issues and the project decisions, and to design the solution. In evaluating the business processes, it is helpful to do the following:

IT touch-points—If it has not been recently done, you should document your IT organization and infrastructure. Next, you should identify needed integration to existing systems, data storage, training, and IT support. The biometric system you choose should scale to enterprise-wide level, as needed. Where are the bottlenecks? What are the pain points? How can you improve that? Will you need to augment your staff to support the biometric system initiative? If so, by what point in the process?

Process change—Make some assumptions regarding how the biometric solution can affect your business processes. What new information will it provide? Establish your written baseline process. How can you improve the current processes? How does automation of processes help? What new metrics are needed to track that improvement? What new information will you receive that will be useful?

Data collection and reporting—Now you need to determine how you will collect and report the information. If you need to support multiple locations collecting and processing data, you may need to consider a distributed configuration. Biometrics can be a good point solution, but it is a better system solution. Data collection across the entire system needs to be considered. Moreover, it is important to define the types of information to be collected and with whom it will be shared. The latter should be determined in consideration of privacy concerns.

Privacy concerns—Anticipate privacy issues. Take steps to mitigate privacy intrusion issues to the extent that they apply to your business and to the application being addressed. Ensure your system controls the collection, use, and release of personal information, and that it protects that information against all forms of unauthorized disclosures. Make a full disclosure of your biometric initiatives to your employees and to your customers. Then ensure that you widely communicate what steps you are taking to protect everyone's privacy as well as what benefits the biometric solution is providing.

Contingency planning—There must be a plan to resolve any failure to enroll or failure to acquire situations, as well as false rejections. It is very important to have a predetermined contingency procedure backed by established policy to handle these common situations. The contingency plan is often a balancing act between user convenience and enterprise security. Know under what conditions your enterprise will implement alternatives for individuals who cannot use the biometric system.

Enrollment—The design of a biometric system should include the following:

- Establish a secure authentication process that enables users to prove their identities via multiple, high quality credentials.
- Provide extensive training and rehearsals for those who will manage the enrollment process so that they understand their duties, their roles, the function of the biometrics, and the importance of enrollment. For many enrollees, this will be their first exposure to a biometric system. The enrollment managers must give proper instruction and exhibit their acceptance of criticism and other forms of enrollee feedback.

- Design security controls that provide only authorized viewers access to the enrollment procedures.
- Select policies and procedures for monitoring the use of the biometrics and for protecting personal information, including rendering decisions on the storage of the biometric templates and all personal identifiers. Know what personal information to store and where to store it, and when to delete it.
- Establish procedures for in-person, secure enrollment activities, including retesting as necessary or implementing alternatives for FTE enrollees. This must include the design and location of enrollment stations as well as the following:
  - Enrollment overview—A description of the enrollment process as well as comprehensive enrollment directions.
  - Enrollment instructions—A detailed step-by-step script for executing the enrollment.
  - Verification overview—A description of the verification process and comprehensive verification directions.
  - Verification instructions—A detailed step-by-step script for executing the verification transaction.
- Plan a seamless, secure, and professional enrollment so that the enrollees do not experience inconvenience or intrusive behavior. The enrollment process should be straightforward, must gather quality data quickly and accurately, and should provide a frustration-free environment. Where possible, educate the users regarding its purpose, the process, and privacy protections associated with the biometric system.

Better information enables process improvements; better processes provide more accurate and timelier information. However, improved processes bring their own challenges. You should accept that integrating a biometric solution might be disruptive to your current processes, but understand that the new biometric system will provide an overall benefit that is compelling. For a successful implementation, your company must be willing to not only adopt the biometric technology but to improve upon the legacy business processes that may be impacted by the biometric solution. It is essential that biometric technologies be molded to fit your business processes, and those business processes must be agile and robust enough to accommodate the changes brought about through the biometric technology, or they should be reengineered.

TEST THE SOLUTION

As with other important events in your life, you should practice, practice some more, and then have a dress rehearsal before the debut of your biometric system pilot. You need to test the biometric readers and then the entire end-to-end system, including enrollment training and reporting. There are two basic yet critical progressions of testing: (1) the walk-through and (2) system testing.

The Walk-Through

The walk-through will help validate where you will perform initial and ongoing enrollments, and where each biometric reader will be positioned. You will want to ensure that adequate power outlets are near to the locations where you need them. It is a good practice to document the physical layout of all system components and power outlets.

System Testing

System testing is a step-by-step review of each component of the entire system, as well as a set of comprehensive end-to-end tests. This is effectively an operational evaluation (discussed in Chapter 6). The purpose of the testing is to ensure that the biometric system functions as advertised and to show the supporting cast (e.g., trainers, IT, security) their integral roles in your system, and to evaluate the workflow and the procedures that your organization will follow. For this you will need a moderate number of volunteers. The components you selected should be proven functional and compatible. It is usually appropriate to document the results of the evaluation, including an initial set of performance metrics: false acceptance rate (FAR), false rejection rate (FRR), and failure to enroll (FTE), as well as formal feedback from the tested participants.

DEPLOY THE SOLUTION

When you actually pilot your biometric solution, you are conducting a real-world lab test. The biometric system pilot validates the concept vis-à-vis your project objectives. It is also in the pilot where issues that did not reveal themselves in previous tests suddenly surface. This is a unique chance to find them and expel them quickly before they can mount a serious concern.

As you prepare to launch your pilot, there are some fairly basic support services that you must address. In some organizations the pilot project manager may have to supervise these support services; in large organizations, there are usually other managers to whom you can delegate these activities. The support services are:

Equipment procurement—Biometric components, including scanners, enrollment devices, and application software must be in place. You may need additional electrical outlets or you might extend some electrical wiring to the locations where the equipment will be used. You might require an uninterruptible power supply (UPS) if you don't already have one. You may need to order some additional PCs or servers.

Application support—You will want to finalize who is providing your application support. What portion is coming from your vendor and what portion from your IT department or other group? Will you require any access to a Web-based portal? If so, who is supporting that effort?

Staff training—You will want to provide your staff with an overview of the technology and to train your staff on how the biometric system may affect their daily procedures. Discuss in detail changes in those procedures.

System documentation—You will need to provision for documenting all aspects of your new system. What happens if problems surface? You will need to document work around procedures and communicate those concepts to the staff. It might be prudent to create an escalation strategy by which a problem not solved by one individual or group is then passed on to the next, which has more insight into a lower level aspect of it.

Monitoring the network—You will want to network your biometric scanners and any peripheral devices so that you can dynamically access and manage them. Generally, you will need to develop or acquire a monitoring system that can do the following:

- Remotely monitor the overall performance and status of all devices connected to the network.
- Provide detailed performance and status information on the various network devices.
- Provide a basic level of alerts, indicating that one or more biometric scanners or other devices is malfunctioning.
- Enable remote system configuration.

- Manage maintenance information including access to the maintenance history for each reader device on the network.
- Generate online reports regarding the status of the network, both predesigned reports with specified periodicity and ad hoc reports.

Maintain metrics measurements—Nothing helps a project manager to quantify the success of a pilot more than metric measurements, and nothing provides better warning to a project manager when things start to go badly. It is often pragmatic to maintain the data in specific time intervals. The following are some basic metrics used in the biometric industry to measure network availability and health:

- *Total use volume* provides a level set for the size of the biometric system It is the total biometric transactions over a given period of time, usually a day or month.
- *Average use volume* is a measurement of the average number of biometric transactions by hour or by day. This metric will identify the time frames for the greatest biometric activity and enable the organization to avoid bottlenecks and other queues.
- *FTE/FTA* rates provide usability metrics that will reveal the degree to which the users have difficulties with the selected biometric solution.
- *Transaction duration* is the time it takes for an individual to submit his finger or hand for verification until an approval or denial is given. This metric provides insight into whether additional training is needed.
- *FAR/FRR* is important to track the continuity of the rates for each biometric scanner. For example, if the FRR is averaging .01% every month at a given reader and then suddenly shows .9%, there may be an equipment issue.
- *Mean time between failure for scanners* is a conventional engineering metric for electronic equipment.

Communicate results—Whether you review the technical, financial, or operational aspects of the biometric pilot, a successful project manager communicates results. You want to quantify not only the metric measurements but also the business benefits that you are receiving. You probably explained it all to your organization when you wrote your plan, but now you are in a pilot and you need to quantify all the process improvements that you are experiencing. Before, you were just talking about it; now you are actually doing it! Educate your entire enterprise. Review the benefits you have attained, including the return on investment, and share that information.

EVALUATE THE PROJECT

Whether the project was an overwhelming success or it had some aspects that performed poorly, it is important to perform an evaluation. A thorough evaluation will provide a basis upon which you can improve in your next implementation. It will also reveal what you did well and what tasks need improvement. Finally, it will provide closure for this specific project.

Any new biometric system is going to experience some issues. Do not hide from them. Aggressively track, record, and resolve problems. You want to detect anomalies before they become problems. Build exception routines for potential failures. Moreover, you need to continually test your entire biometric system, including back-end applications.

Much of the rationale for the vein pattern biometric pilot was probably to improve security and efficiency in a privacy-enhancing way. Evaluate over time how well that has occurred. Address any inconsistencies or areas that do not appear to have improved. Even if the biometric pilot is an overwhelming success, you need to evaluate ways in which you can make your procedures even more efficient. You want to future-proof your system by looking at all the ways it might be impacted by new developments or unfortunate events.

Another technique to evaluate a completed project is a lessons-learned session. This is a meeting of the entire project team. Its purpose is to document the lessons learned for use in future biometric projects. This is especially helpful if your organization has adopted a phased approach to implementations.

Finally, all projects should include a written summary report authored by the project manager and submitted to senior management. This is sort of a final status report that summarizes the entire project. The final status report should summarize the project background as well as the goal, objectives, and scope. It should compare the final results to the baseline schedule, explain major discrepancies, and include any results of surveys conducted about the project and any lessons learned. It is considered good form to include the final Gantt chart describing the actual progress of the biometric system project.

An Implementation Road Map

In full production, the biometric network may grow much larger than it was in the pilot. As more biometric readers and users are added, the system will become much more complex. As you move from pilots to large-scale deployments, you will need biometric system infrastructures that are scalable and solutions that are flexible. This is where you establish a plan from your pilot to your production. A good road map is an integrated plan for achieving the business objectives within a specific timeframe with estimated costs. It should include all the control plans we discussed. Indeed, the pilot could be considered a miniproduction plan. Only now you are experienced as a biometric project manager. You have implemented a system and learned a lot about the biometric technology, but mostly you learned how your organization adapted to it. With all you have learned, your next biometric deployment should be a walk in the park!

9

Issues in Vein Pattern Recognition

This chapter will discuss several commonly cited concerns about biometrics as well as some popular misunderstandings, such as the expectation for biometrics to end identity theft, bring control mechanisms to our national borders, and identify terrorists. Biometric technology holds a lot of promise to improve upon the quality of life of all citizens as well as protect them from a variety of unwarranted intrusions. However, it is not a silver bullet that will provide neat, risk-free solutions to the security threats that are now part of our modern world. What biometric technology can do is provide a strong authentication component to security solutions without compromising privacy or increasing costs beyond the value that they provide. This chapter addresses the two key concerns associated with the deployment of vein pattern recognition (VPR) technology: privacy and the business case.

PRIVACY

For several decades commercial enterprises of all types have been gathering personally identifiable data on consumers for use in direct marketing and for other purposes. Citizens throughout the world are becoming highly sensitized about the quantity and type of personal data that are being collected about them,* and then used by others. New technologies

* Personal data is data that relates to an identifiable individual. The International Biometric Industry Association (IBIA), the U.S. biometric trade association, has declared, "Biometric data is electronic code that is separate and distinct from personal data."

enable data collection to accelerate, and to connect to other depositories of data. It is natural that biometrics would open a whole new venue of concern. Misinformation and sensationalized stories of privacy abuse tend to heighten the public's fear and distrust. This is a great concern to the biometric industry because threats to privacy and fear of the technology will certainly inhibit user acceptance of biometrics.

Let us be frank about this. Privacy is a very subjective notion. For some people, those who just want to be left alone, the receipt of unwanted mail is a privacy intrusion. Many regard information about themselves as private property and do not want it known to anyone that they have not personally told. To better appreciate the size of this group, just consider the growing propensity toward unlisted phone numbers. Others view most of their personal information as publicly available data, or they simply do not ponder the privacy implications.

Privacy Threats

For most of us, privacy is basically a policy matter. Although biometric technology can enhance and support most commonly held notions of privacy, it cannot guarantee it. Clearly, the topic of biometrics often raises the public's angst regarding privacy. There has long existed within some groups an inherent discomfort with biometrics; they have a perception that biometrics represents privacy intrusiveness. As mentioned in an earlier chapter, we leave our fingerprints on everything we touch, and there are limited prohibitions against someone taking our photographs or recording our voices, each of which can be accomplished legally without our awareness or consent, dependent on the circumstances. However, most privacy concerns regarding biometrics generally fall into one of the following seven scenarios:

1. *Creation of a complete profile on individuals*—Any identification system, whether based on biometrics or something else, that establishes a database with unique identifiers bound to personal information will generate a series of privacy concerns. The public clearly worries about unique identifiers where biometric data connects otherwise disparate information about a person. The biggest concern is the secondary use of the information linked to the unique identifier. The issue appears to be the potential that personal data, no matter how diverse, can somehow be connected to form a personal dossier using the unique biometric identifier.

There is the possibility that the established database can be hacked or a database administrator can gain unlawful access to that data.

2. *Information access without consent*—The potential for third-party use of biometric data is a privacy threat because it erodes the personal control of an individual over his information. This is really about freedom of choice where an individual should be able to choose to provide his biometric data or not, and about self-determination regarding to whom he wishes to give it. Third-party access to biometric information without the consent of the individual is a clear privacy violation, as individuals should have control over their personal information.
3. *Use of information for unintended purposes*—Many people worry about the expansion of uses of biometric data, a sort of function creep. They fear that biometric data collected for one's driver's license or passport might be used to track down people who do not pay their taxes on time or who have outstanding minor civil violations. This anxiety becomes more acute when it is applied to a government agency or an enterprise that may want to track behavior of an individual or group of individuals. This perception may not be far off target given the extent to which governments and corporations maintain huge repositories of data on the average citizen, albeit with mostly benign intent. If a database contains personally identifiable information or if it is part of a critical security system, then it may become a target for attack by any number of malicious individuals or nefarious organizations whose goals are to compromise such installations. Additionally, there are concerns that over time, a specific, unique biometric identifier could cross all applications and subsequently be used without proper oversight by law enforcement agencies.
4. *Concern for the proportionality principle*—Some people view the use of biometric identifiers as overkill for many applications in everyday life. They cite the original employment of biometrics as a technology reserved for high security (e.g., military applications) or high-risk individuals (e.g., criminals). Today, biometric systems are used in novel and creative ways for emerging applications that include banking and shopping. The issue of proportionality has now surfaced: what proportion of an individual's private information should be required for the purpose of access or service. This has come to be referred to as the proportionality

or reasonableness test. Is it reasonable that a person must provide a biometric identifier to use his debit or credit card or to buy a pack of cigarettes (i.e., to prove age)?

5. *Surveillance and profiling*[*]—Use of biometric techniques for surveillance, especially covert surveillance, is certainly a valid privacy concern. Individuals have no control if information about them is collected or used to track them without their consent. Most people do not want to become part of a surveillance database. Moreover, a corollary concern is the potential for profiling that real-time surveillance provides.
6. *Compromise of biometric identifiers*—The fact that biometrics is a powerful identity verification tool means that the biometric characteristic cannot be easily changed if a biometric becomes compromised. It may be difficult to establish proof of misuse by imposters. The possibility that a biometric template could be stolen or compromised raises serious concerns. However, it must be kept in mind that a biometric solution matches a template from a "live sample" with a reference template at the point of transaction; the use of a live sample somewhat assuages this concern.
7. *The loss of anonymity*—By its very nature, biometrics is not in alignment with the concept of anonymity. As biometrics continues to grow and both public and private sectors continue to widen its usage, some individuals may be asked to produce a biometric identifier in quite unexpected and potentially unwelcomed situations. However, most people believe that the individual should have sole authority over the distribution of his or her biometric information. Further, anonymity is frequently linked to autonomy, an individual's sense of freedom and independence. The continued erosion of anonymity may impact people's views of their individual autonomy.

At the vortex of most privacy issues are concerns about the potential misuse of biometrics, such as a biometric system pointing to other information pertaining to an individual. Thus, an intrusion on one's privacy tends to originate from the inherent misuse of the technology, not from the biometric application itself. Authentication applications may be less threatening to privacy simply because the individual knowingly enrolls

[*] Profiling refers to recording an individual's behavior and analyzing psychological characteristics in order to assess or predict his actions in a given situation or to identify a particular subgroup of people.

in a given system and thereby gives permission to use his or her biometric characteristics for a specific purpose. However, that information can still be misused.

It is generally the case that where there is biometric information, there is probably some personal information. An enterprise planning to use biometrics should conduct a privacy assessment to analyze what personal information is being collected and the necessity for doing so, as well as how that personal information will be used. The enterprise should carefully examine data storage. For example, the use of the smart card increases security and enables strong authentication in an off-line environment. Further, matching algorithms can be implemented directly via on-card matching. In this way, the enrolled biometric template never leaves the card for any reason. And the individual controls and carries the card. This is certainly a privacy-enhancing method of storing one's biometric identifiers.

Another privacy protection is the practice of not storing raw biometric data. Although biometric vendors insist that the original biometric trait cannot be restored from an enrollment template, how does an average citizen confirm that? The algorithms used in biometric systems tend to be proprietary, and they are usually not available to public scrutiny. When an individual no longer participates in a biometric application, is all his personal information expunged from the system's data repository? These are very real issues that the public wants addressed in order to become more comfortable with the use of biometrics.

Biometrics can be privacy threatening if safeguards are not designed into biometric systems from the beginning. It really depends on how the biometric system is deployed, how the biometric templates are stored, and the strength and vigilance of the safeguards that are put in place to guard against third party access or linking inappropriately to other systems.

Some in the biometrics industry do not have a comprehensive understanding of the underlying privacy issues related to biometrics. Many privacy advocates present very cogent concerns about potential abuse of biometrics. On the other hand, there are some privacy zealots who do not possess adequate knowledge of the current state of biometric technology such that they can fully weigh the real against the fictional. Their credibility is undermined when they suggest scenarios that are incompatible with the basic capabilities and limitations of biometric technology.

Regarding privacy, not all biometric systems or their applications should be painted with the same brush. There are some truly privacy-intrusive biometric systems such as the ones that are deployed without our knowledge or consent. They may be used for undisclosed purposes. Of course, some of these privacy-threatening systems are used by national governments for surveillance purposes and come under the banner of "national security" or better law enforcement; others are simply ill designed or inappropriately deployed.

Privacy Enhancements

There is no valid reason for biometric technology to be privacy threatening. Biometrics can be used to enhance privacy, and there are many ways to accomplish this. There are many real advantages to being able to prove definitively that you are who you say you are. Moreover, biometrics can help protect other types of personal data in the context of data security by providing a stronger personal binding of access rights. For example, using a biometric feature to encrypt security keys could enable access to a bank account with heightened security while improving user convenience.

Biometrics can provide an individual with personal control over who has access to his information. The employment of biometrics is one of the best ways to "lock up" one's records or to ensure that one's identity cannot be stolen. Biometrics can enable authentication without necessarily revealing a person's identity. For example, there are many situations in which someone needs to authenticate age but not identity. In addition, by using multiple biometric features, it is possible to segregate related personal information and thus limit the erosion of privacy through the linkage of separate sets of data.

The judicial use of biometrics can preserve one's privacy in several ways: First, by securely verifying the identity of an individual, biometrics protects that identity from fraud. Second, biometrics should provide *only* the type of personal data that is needed by a given authority for a specific application. A typical example is a driver's license or identification card that contains one's address, age, visual impairments, and so forth. Not all that information needs to be provided to a hotel clerk who simply wants to verify name and address. Third, technology advancements enable only biometric templates to be stored, not the actual biometric raw material. This is very advantageous in that it somewhat assuages some privacy concerns about storing biometric identifiers, and it requires less memory and less processing power, reducing the cost of the biometric system.

Not only is it unlikely that one can "reverse engineer" a sample from a template, in some cases biometric systems can actually prevent it. Additionally, vendor templates are not generally interchangeable. That means that templates from one vendor cannot be compared or tracked against another vendor's set of templates. The data formats are proprietary and therefore incompatible. Moreover, virtually all commercial biometric systems encrypt their templates.

There has emerged in the past decade the concept of privacy-enhancing technologies (PETs). It is an umbrella term for a range of beneficial technologies with a common function: to protect sensitive personal data. One might use a biometric for access to a personal item such as a personal digital assistant (PDA), laptop computer, or home access control. Biometric systems that use specific design elements to ensure that biometric data is protected from unauthorized access and inappropriate usage are considered *privacy sympathetic*. Most business applications can incorporate privacy-sympathetic elements into their biometric systems. Those systems that use biometrics to protect or limit access to personal information or provide a means for an individual to establish a trusted identity are referred to as *privacy protective*. Use of biometrics at automated teller machines (ATMs) to protect one's financial information is an example of a privacy-protecting biometric system.

In the context of biometric data, the aim of PETs is to protect one's privacy by eliminating or reducing the storage and transfer of personal data without loss of function. Of course, some technologies are considered *privacy neutral* in that these technologies neither protect nor threaten privacy; and some are considered *privacy invasive* as they enable access and usage of personal data in a manner inconsistent with generally accepted privacy principles.

The manner in which a biometric system is deployed can significantly impact its ability to protect or threaten the privacy of its users. Of course, the use of biometrics should conform to the privacy requirements of the enterprise or government implementing it. But there should be room for privacy adjustments. For example, is it always necessary to know someone's identity separate from an authentication of his or her credentials to grant access privileges? In many scenarios, the identity is not needed, just the credential validation. For those situations in which someone must surrender his or her identity, the enterprise might enact PET rules similar to the following:

- Use as little personal data as necessary for the authentication.
- Control the collection, use, and release of personal information.
- Encrypt the personal data and safeguard personal information from disclosure to third parties.

- Delete the personal identifiable data as soon as possible.
- When not absolutely required, do not use central databases.
- Give users control over personal information.
- Provide a guaranteed level of trust by using evaluation and certification.

Privacy Legislation

Personal privacy and its protection are essential concerns of all democratic countries. It is a cornerstone issue essential to the autonomy of the individual. Some have argued that privacy is a fundamental tenet of human dignity. As a result there has been increased regulatory focus on privacy issues globally, especially in the past 20 years. Technological developments such as the Internet, contactless payments, 3G mobile phones, and the ascent of biometrics have altered the way that people conduct business, often with increased risks to privacy. The heightened level of legislative activity in Europe, North America, and Asia, among other regions, has manifested itself in a series of sweeping legislation.

Europeans are acutely sensitized to privacy issues. After unfortunate experiences with the communist party in Eastern Europe and the WWII-era fascist governments, most Europeans look unfavorably at unbridled use of personal information. Europeans have a healthy distrust of corporations and governments when it comes to potential for abuse of personal information. The right to privacy is a key tenet of European laws. In 1995 the European Union Data Protection Directive, also known as the European Union Directive 95/46/EC, required all European Union member countries to adopt national regulations to standardize the protection of data privacy for citizens. This directive focused on the processing of personal data and on the free movement of such data.

Each European member country had to comply with the directive by enacting its privacy provisions. For some countries, such as the United Kingdom, this required new legislation. The United Kingdom Parliament passed the Data Protection Act in 1998 and thereby made new mandatory provisions for the regulation of the processing of information relating to individuals. The act created rights for those who have their data stored, and it imposed responsibilities and obligations for those who collect, store, or use personal information. Although this act did not directly refer to privacy, it defined a legal framework that governed the protection of personal data in the UK. The act defined eight "data protection principles" of information handling and covered any data (with some notable exceptions, such as national security) that could be used to identify a living person, including names, phone numbers, and e-mail addresses, and which was held or intended to be held on computers or a "relevant filing system."

Canada has long been a leader in privacy legislation. Canadian law regulates the type of information that governments can collect, how it can be used, and how citizens can access, challenge, and amend the information as codified in its 1983 Privacy Act. The personal Information Protection and Electronic Documents Act provides legal guidelines for the collection, storage, and use of personal information by a variety of organizations in the private sector. The Act is overseen by the Office of the Privacy Commissioner.

So far the United States has not relied on federal regulations for personal data protection or information privacy in the private sector. Indeed, the United States lacks any single overarching legislation that is directly comparable to the EU Directive. Instead, the U.S. approach has been to coerce corporations to self-regulate and to pass legislation when needed to address specific circumstances such as the Children's Online Privacy Protection Act (COPPA), which prohibits commercial Web sites from collecting information from children; the Health Insurance Portability and Accountability Act (HIPAA) for privacy and security of health care information; the Gramm–Leach–Bliley Act for private financial information; and the Fair Credit Reporting Act, which protects personal credit worthiness. Each of these specific legislative acts has comprehensively addressed personal information and privacy; but it does so selectively, not in an overarching legislative mode.

Even the generic sounding Privacy Act of 1974 (PA74) does not really protect American citizens' informational privacy. Passed in response to privacy abuses that ran rampant in the Nixon Administration, PA74 regulates the collection, maintenance, use, storage, and distribution of personal information by the executive branch agencies. It codified disclosure prohibitions and agency recordkeeping requirements for "fair information practices." Later, the Computer Matching and Privacy Protection Act of 1988 updated PA74 with regard to automated matching programs. However, the private information held by the courts, corporations, and even nonagency government entities are not subject to the provisions of either act.

Japan's Personal Information and Protection Act (PIPA) took effect on April 1, 2005. Not unlike the United States, Japan has long preferred encouragement of industry self-regulation over national legislation. However, throughout the 1990s and into the 2000s, Japanese media carried increasingly disturbing reports of privacy abuses in Japan by businesses and the local and national governments. After a series of high-profile scandals regarding confidential information in 2002 and 2003, the Japanese government felt that it had to do something. On the one hand, it was important for Japan to preserve the utility of information technology and not handicap its industries; but on the other hand, the government heard the plea of its citizenry to protect personal information from misuse and unauthorized disclosure.

Similar to the United Kingdom's Data Protection Act, PIPA defines personal information or personal data to mean any information that can be used to identity any living person.* PIPA applies to all personal information handlers (PIHs),† and it imposes an array of restrictions relating to how PIHs treat personal information. Effectively, PIPA accomplishes three key things: (1) it regulates the handling of private information by industry regarding living individuals including the requirement to specify a "purpose of use" to process personal information; (2) it establishes rules of conduct in regard to how personal information will be acquired, used, maintained, and safeguarded while providing individuals access to this information; and (3) it imposes penalties for failure to adhere to the prescribed standards and rules of conduct.

The more policy measures are able to encourage the use of biometrics to enhance privacy, the more biometrics will be acceptable to the public at large. It could be argued that all biometric systems should be designed to protect personal data. There are many ways to do this.

Balancing Privacy and Security

Humans have long recognized one another through their physical and behavioral traits: facial features of a colleague, a leader's gait, or an actor's voice. Now we use automated technology and defer recognition to a computer. Why? Because we need a foolproof means of identity verification for people we do not know and may never meet. Biometric technology has been designed for remote use in an unsupervised environment where it is critical to sort out friend from foe.

We have already tried personal identification numbers (PINs), passwords, keys, and proximity cards—all of which we forget, lose, or

* PIPA provides no protection to corporations, businesses, or associations.

† PIHs are those that manage a database of more than 5,000 individuals to conduct business.

compromise, rendering them insecure and, by extension, unreliable. Our society relies more and more on remote electronic communications and e-commerce, and such transactions need the accompaniment of remote authentication. Our Web-enabled world demands reliable user authentication. Consider the consequences of unsecure access systems in our enterprise environment with the potential for disclosure of sensitive, confidential information; compromised systems; and denial of service.

The truth is that it is getting progressively harder to preserve our privacy *without* employing biometrics or something like it. Traditionally, data security has relied on something you know (user name and password) to grant information access. Unfortunately, single sign-on not withstanding, password management has become increasingly complex and dynamic (such as when well-meaning IT [information technology] staff forces new passwords each month or each week on enterprise employees). So it is not surprising that users resort to poor security practices (such as writing down their passwords) to combat their inability to remember many diverse ID–password combinations. For at least the past four decades, physical security has relied on what you have (e.g., badges, keys, or access cards). These solutions have proven relatively ineffective, as they are easily lost, replicated, or stolen. Because of these limitations, solutions that rely on "who you are" easily trump the "what you have" solutions.

In our zeal to embrace biometrics, we must be cognizant that this new technology must be more reliable, more user friendly, and more manageable than what it is replacing. And it needs to do one additional thing: preserve privacy. Privacy must be central to the deployment of biometrics in either the public or private sectors. To ensure that biometric deployments are in full alignment with generally accepted privacy principles, particular attention must be paid to:

- Data collection, storage, and retention
- Safeguarding biometric identifiers with strict physical and logical access controls
- Obtaining individual consent over the information collected

Best Practices

Privacy considerations are key to a well-designed and effective biometric system. To be perceived as privacy enhancing, a biometric system should fully satisfy a set of reasonable privacy directives. Several government organizations and industry groups have made recommendations to safeguard privacy in the deployment of biometrics. In 2003, the International Biometric Group (IBG) published a set of practices in which it suggested general privacy principles. These best practices were meant to serve as guidelines for privacy-sympathetic and privacy-protective deployments. Enterprises may not be able to adhere to all these practices, but they are encouraged to implement as many as possible without undermining the basic operation of their biometric systems. The following are a summarized version of IBG's bioprivacy best practices:*

Scope and Capabilities

- *Limit system scope*—Biometrics should be collected for a specific stated purpose.
- *Limit retention of biometric information*—Although enrollment data is stored in most systems, verification data can usually be discarded.
- *Efficiently manage system storage of identifiable biometric data*—Any actual images, recordings, and identifiable biometric data should be discarded as soon as possible.
- *Limit collection, storage of extraneous information*—Collect nonbiometric data only as absolutely necessary.
- *Make provisions for system termination*—Establish a policy to depopulate and dismantle the system.

Data Protection

- *Use security tools to protect biometric information*—Use encryption, private networks, and secure facilities to protect biometric information at all stages of its life cycle.
- *Protect postmatch decisions*—Protect match, nonmatch, and error data transmissions.
- *Limit system access*—Prevent internal compromise of the biometric data by limiting access to only a specific group of system operators.

* For a comprehensive list of the IBG's BioPrivacy Best Practices, visit www.biometricgroup.com/reports/public/reports/privacy_best_practices.html.

User Control of Personal Data

- *Make system usage voluntary when possible and allow for unenrollment*—Provide an opt-out mechanism for those who do not wish to participate.
- *Allow correction and access of biometric-related information*—Enable all users to view, correct, and update their information stored in a biometric system and to re-enroll, if necessary.

Disclosure, Auditing, and Accountability

- *Disclose system purpose and objectives*—Explain the purpose of the biometric system to both operators and enrollees. Disclose whether enrollment is mandatory or opt-in and describe the fallback procedure.
- *Hold operators accountable for system misuse*—Disclose who is responsible for the biometric system and provide for a means of dispute resolution.
- *Disclose the use of the biometric system during enrollment*—Full disclosure is the key to user buy-in.
- *Make provisions for third-party auditing and oversight*—Make provisions for operational oversight and review which are critical to all biometric systems.

THE BUSINESS CASE

Widespread security concerns such as identity theft and terrorist attacks have acquired a national spotlight throughout this decade. Each sensational news story of identity theft and each television broadcast on terrorist activities or even suspected activities tend to turboboost the general interest of the public in biometric technology. However, sometimes publicity is not a positive thing. Unrealistic claims of biometric system performance engender unrealistic expectations regarding the capability and use of biometric systems. For example, biometrics can certainly assist in combating identity theft, but it does not provide a comprehensive solution as a standalone technology. Biometrics does not yet provide a foolproof method to determine who performed a given action or who visited a given venue.*

* See "The Use of Technology to Combat Identity Theft; Report on the Study Conducted Pursuant to Section 157 of the Fair and Accurate Credit Transactions Act of 2003," February 2005, U.S. Department of the Treasury.

Biometrics remains part of an overall security solution, and it is only as effective as the system in its entirety.

The public's awareness of and familiarity with biometrics is rising quickly. It seems that technology familiarity can transform into an acceptable comfort level. As people gain a better understanding of biometric technology, they fear it less, and they better acquaint themselves with its promise as well as its limitations. Biometrics remains one of a growing set of tools that government, enterprises, and even individuals can use to combat security concerns. But quantifying the value of security in terms of return on investment is a bit problematic.

For many enterprises, higher security is a priority. Enterprises of all sizes exhibit concerns about security as each must address the risks of security breaches. As previously stated, biometric implementations exist as part of the overall security management infrastructure. Each organization must determine its security investment level vis-à-vis the value of the assets that need protection, including its staff and customers. Given that security investments tend to have diminishing returns as the investment increases, assessing how much security is enough is one of the key challenges faced by most enterprises. It is also important to carefully select the biometric system that can achieve the right fit for the enterprise. However, a reduction of fraud generally results in significant benefits to an enterprise, justifying the investment in biometric technology.

Various governments at the national, state/province, and even local levels are encouraging improved physical and informational security. Some governments are legislating security solutions, requiring compliance for access controls, audits, and privacy protections. Therefore, the risk analysis of any enterprise must examine multiple dimensions of the risk that inadequate security enables, including penalties for not complying with regulations and legislative directives, bad press caused by shoddy security, harm caused to the enterprise, and the opportunity costs of not solving one's business problems.

Let us address a problem common to every company—passwords. Improved security might achieve a partial investment return by reducing the growing costs of managing passwords. As networks and applications grow, issues of password management are increasing at an accelerating rate. Not only is there lost productivity for those who forget their passwords and are temporarily locked out of their PC or network, but the costs incurred in resetting them via an IT department or help desk intervention can be astounding. Moreover, employees who fail to comply with password policies further increase the vulnerability of the enterprise. Elimination of these numerous yet unnecessary password issues is a solid basis upon which to build a return on investment for a biometric-based security system.

The cost of biometric authentication is not inexpensive, but it is affordable. There are costs for the hardware, the application software, maintenance costs, and integration to back-office security systems. However, advancements in accuracy and scalability render biometric technology a viable solution for many applications.

As is the case with most technologies, costs for biometric systems should continue to fall as the technology advances and its uses grow. Indeed, the cost of a biometric implementation is no longer a major obstacle in many buying decisions. Although the price–return ratio is on a downward trek, it may still not be low enough for some applications. To make that determination, one must identify the tangible and intangible benefits, including user convenience, security, flexibility, usability, privacy, and high-tech image, and then compare them to the initial investment and ongoing costs of a biometric system.

Biometrics is effectively an all-or-nothing technology. That means that if a biometric system is deployed, it must cover everyone within a given enterprise or a given application. If users are accessing the network through a biometric authentication system, then a user who is not using biometrics would tend to undermine the system. Said differently, a user who could not use biometric authentication would have to have an alternative means to authenticate to the network, and the existence of many such alternatives increases an enterprise's overhead and its vulnerabilities.

What is driving biometrics? Biometrics is being driven by the usual push–pull forces that propel any technology to prominence. The market push is the ever-present need for stronger security as well as a growing fear of fraud and the threat of identity theft as fueled by high-profile security breaches. Other authentication methods—PINs, passwords, badges, and other tokens—cannot assure the identity of the user. Accompanying that fear and concern are decreasing infrastructure costs to implement biometrics and an increasing number of successful pilots that serve as reference implementations. Additionally, government laws and policies, and network-based enterprise applications demand heightened security.

The market pull is that exciting new technologies enable new business applications, and this creates new business opportunities. The public sector has been introducing biometrically enabled passports and national ID cards, government health cards, social security cards, and driver's licenses as well as new welfare support instruments. In the commercial sector, financial services are developing new personalized, biometric-enabled services. Biometrics as a tool can actually increase customer satisfaction and improve customer relations. Biometrics helps to improve competitiveness for many players in this large industry sector. Single sign-on has been with us for quite a while, but it still consists of at least one set of difficult-to-remember ID–password combinations and often still cannot definitively verify identity. On the other hand, biometric enablement provides true single sign-on.

Biometric-based identity verification accomplishes the following:

- Combats fraud.
- Improves services by providing a consistent method for verifying identity.
- Provides a viable platform for electronic service delivery.
- Enhances convenience and reduces costs for many applications.
- Provides confidence in the integrity of benefit programs.
- Improves access to eligibility information.
- Reduces some administration costs with improved data quality and processing time.
- Diminishes redundant data collection.
- Lessens operational costs.

In the United States, heightened security and information protection are being legislated along with biometric verification. The Truck and Bus Safety and Regulatory Reform Act of 1988 required "minimum uniform standards for the biometric identification of commercial drivers." The Illegal Immigration Reform and Immigrant Responsibility Act of 1996 requires the Immigration and Naturalization Service (INS) to place on its border crossing cards "a biometric identifier (such as the fingerprint or handprint of an alien) that is machine readable." To ensure confidentiality and data integrity, HIPAA strongly encourages the use of technology to support access control of sensitive health care information, including data encryption when that information is transmitted across telecommunications channels. The Patriot Act specifically required the U.S. federal government to "develop and certify a technology standard that can be used to verify the identity of persons" seeking entry into the United States.* The Enhanced Border Security and Visa Entry Reform Act of

* Patriot Act, Section 403(c).

2002 requires "machine-readable, tamper-resistant visas and other travel and entry documents that use biometric identifiers" be issued to aliens.* The U.S. Visitor and Immigrant Status Indicator Technology (US-VISIT) program enhances control at U.S. borders by verifying the identities of alien visitors with visas via biometrics. Several of these acts require the National Institute of Standards and Technology (NIST) to develop certification standards for determining the accuracy of some selected biometric technologies.

Biometric solutions have many opportunities for growth. The technology will continue to improve and will resolve the many challenges it faces. Interoperability will be solved through standardization, which will lead to even greater market adoption and acceptance. As the technology continues to improve and as costs inevitably decrease, the demand for biometrically enabled solutions will increase along with the need for greater security and greater convenience.

SUMMARY

Biometrics can provide the connectivity between a person and his identity by linking to his measurable characteristics. Both the biometric authentication and identification markets are expanding at an increasing rate. One reason is that awareness of information security has grown significantly over the past decade. Biometric techniques are improving in accuracy and reliability. New and exciting biometric methods are being introduced and piloted.

Biometric technologies are rapidly finding acceptance as the foundation for highly secure identification and personal verification solutions. Some governments are now mandating biometrics for passport issuance, citizen identification cards, or for use in physical or logical access to government facilities. Many industry watchers expect local governments to begin issuing secure ID cards with biometric applications. Increasingly, biometrics is being used in commercial applications such as electronic medical records management, and hotel and apartment access.

The use of biometric technology by government, corporate enterprises, and individual citizens will continue to grow and will do so on an accelerating pace for the next few decades. One could argue that any interface between man and machine is a candidate for biometrics.

* Enhanced Border Security and Visa Entry Reform Act of 2002, Section 303(b)(1).

And as long as the biometric systems deployed have privacy designed into their data collection, storage, and retention, and it is accomplished openly with individual consent, then biometrics can remain privacy enhancing. Additionally, when biometric solutions are properly implemented, they can provide greater security and convenience than alternative technologies.

Vein pattern recognition (VPR) technology is one of the newest, commercially viable biometric modalities to emerge. It has gained traction in multiple venues and applications around the globe. Its relatively small size and multiple form factors render VPR systems convenient, versatile, and low cost. VPR technology is considered robust in handling environmental changes. Moreover, the technology will not function with a severed finger or hand, and it is very difficult to spoof. It is becoming widely deployed in mission-critical applications.

Vein pattern recognition systems can enable an individual to verify his identity, while safeguarding his personal identifier(s). As an identifier, vein patterns are universal, unique, permanent, and usable by 99.98% of virtually any population. The near contactless nature of VPR systems provides a hygienic advantage. Vein patterns are closed view and cannot be easily scanned by surveillance equipment without an individual's awareness and general cooperation. Unlike fingerprints or palm prints, there is no property of latency with vein pattern identifiers. A biometric identifier that is intended for public use should not have forensic properties, and vein pattern recognition technology does not. No manufacturer of vein pattern scanners stores raw biometric data, and all VPR manufacturers encrypt their templates. Vein pattern recognition technology is truly a privacy-enhancing biometric.

GLOSSARY

A

Access: The method of getting to a thing or information, or performing an action.

Access control: The process of granting or denying specific request to enter a physical facility or to access and use information.

Accuracy: A computation indicating a biometric system's ability to correctly score a submitted template as a result of a matching process. FAR, FRR, and EER are computational proxies for accuracy.

Acquisition device: The hardware used to acquire biometric samples.

Active imposter acceptance: Acceptance of a biometric sample submitted by someone attempting to gain illegal entry to a biometric system.

Algorithm: A limited set of well-defined instructions to solve a task. Multiple algorithms are used by biometric systems to perform such tasks as generating a template or determining whether a sample and a template are a match.

Anonymity: The condition of being unknown or unacknowledged; an entity with no known identifier.

Application Program Interface (API): A set of instructions or services used to standardize an application or an application developer tool to build applications. Any system compatible with the API can then be added or interchanged by the application developer.

Application Specific Integrated Circuit (ASIC): An integrated circuit developed for specific applications to improve performance.

Asynchronous multimodality: Biometric systems that require a user to verify through more than one biometric identifier in sequence. Asynchronous multimodal solutions are comprised of two or more distinct authentication processes. A typical user interaction might consist of a verification via finger scan, then via facial recognition, if the finger scan is successful.

Attempt: The submission of a biometric sample to a biometric system for identification or verification. A biometric system may allow more than one attempt to identify or verify.

Audit trail: In computer/network systems, it is a record of events (protocols, written documents, and other evidence) that can be used to trace the activities and usage of a system. Such material is crucial when tracking down successful attacks/attackers, for determining how the attacks happened, and in using this evidence in a court of law.

Authentication: The process of establishing confidence in a given claim. In biometrics, it is any systematic method of confirming the identity of an individual with confidence; it is used interchangeably with *verification.*

Authorization: The administration of person-specific rights, privileges, or access to data or corporate resources.

Automated Fingerprint Identification System (AFIS): A specialized fingerprint system that is used to determine the identity of individuals. It is predominantly used by law enforcement. It automatically matches one or many unknown fingerprints against a database of known prints.

Automatic ID/Auto ID: An umbrella term for any biometric system or other security technology that uses automatic means to check identity. This applies to one-to-one verification and one-to-many identification.

B

Behavioral biometric: A biometric trait or characteristic that is learned and acquired over time rather than a physiological characteristic.

Benchmarking: The process of comparing measured performance against a standard reference.

Bifurcation: A branch made by more than one finger image ridge. It is a Y-shaped split of one ridge into two ridges.

Binning: The process of examining or classifying biometric data in order to accelerate or improve a biometric matching process. This allows a database of biometric data to be presorted to speed up the process of matching captured biometric data with comparison data.

Biometric Application Program Interface (BioAPI): It was designed to produce a standard biometric API aiding developers and consumers. It enables easier installation and integration of biometric devices within the overall system architecture.

Biometric data: The extracted information taken from the biometric sample and used either to build a reference template or to compare against a previously created reference template. The term is sometimes used as a catchall phrase to refer to any data created during a biometric process to include samples, models, templates, or scores. However, it does not include personal information such as user name or demographic information.

Biometric engine: The software element of the biometric system that processes biometric data during the stages of enrollment and capture, extraction, comparison, and matching.

Biometric identifier: An objective measurement of a physiological or behavioral characteristic of an individual that can be used to verify identity or provide an identification if the individual's data is contained in a biometric database.

Biometric sample: The identifiable, unprocessed image or recording (raw data) of a physiological or behavioral characteristic, acquired during submission, and used to generate biometric templates.

Biometric system: An automated system capable of capturing a biometric sample from an end user; extracting biometric data from that sample, comparing the biometric data with that contained in one or more reference templates, deciding how well they match, and indicating whether an identification or verification of identity has been achieved, and storing the biometric information.

Biometric taxonomy: A method of classifying biometrics.

Biometric template: An encoded, formatted, digital representation of an individual's distinct characteristic(s). Templates are typically created by translating an individual's biometric characteristics using sophisticated algorithms. Templates can vary among biometric modalities and vendors.

Biometric throughput: The total time it takes from the capture of a biometric trait to user feedback regarding the outcome.

Biometrics: A technology that uses behavioral or physiological characteristics to automatically determine or verify identity. Also, it is a measurable, physical characteristic or personal behavioral trait used to recognize the identity, or verify the claimed identity, of an enrollee.

Breeder document: The original documentation that is the source of identity to apply for other forms of identity credentials, for example, a birth certificate, driver's license, or passport.

C

Capture: The method of collecting a biometric sample from the end user via a sensor. It might be considered a "live capture" or a "dead capture."

Certification: The process of testing a biometric system to ensure that it meets certain performance criteria. Systems that meet the testing criteria are said to have passed and are certified by the testing organization.

Chain of trust: An attribute of a secure identity system that encompasses all the system's components and processes, and assures that the system as a whole is trustworthy. It guarantees the authenticity of the issuing organizations, devices, equipment, networks, and other components of a secure identification system.

Challenge–response: An automated method to confirm the presence of an authorized person by eliciting direct responses from the individual.

Claim of identity: A claim that occurs when a biometric sample is submitted to a biometric system to verify a claimed identity.

Claimant: A person submitting a biometric sample for verification or identification while claiming a legitimate or false identity.

Closed-set identification: Refers to situations when an unidentified end user is known to be enrolled in the biometric system. It is the opposite of *open-set identification*.

Common Biometric Exchange File Format (CBEFF): A standard for a biometric system to identity and interface with multiple biometric systems, and to exchange data among system components. This specification focuses on the interoperability of biometric-based application programs and systems developed by different vendors by enabling biometric data interchange. CBEFF is being augmented by the NIST/Biometric Consortium Biometric Interoperability, Performance, and Assurance Working Group.

Comparison: The process of comparing a biometric reference or template with a previously stored template.

Contact/contactless: In regard to chip cards: whether the card is read by direct contact with a reader or has a transmitter–receiver system that allows it to be read using radio frequency identification (RFID) technology (up to a certain distance).

Contactless card: Smart cards or memory cards that communicate by a radio signal. The range is normally up to 10 cm from the reader.

Credential: Evidence that attests to one's right to a privilege (e.g., entering a building or crossing an international border); it commonly refers to a physical token such as a badge or electronic information such as a digital certificate.

Credentialing: The processes for creating primary identifying documents, authenticating those documents, and issuing identification documents.

Crossover error rate (CER): Also known as the equal error rate (EER), it is a comparison metric for different biometric devices and technologies. It is defined as the rate at which FAR equals FRR. In general, the lower the CER (or EER), the more accurate the biometric device.

Cumulative match characteristic (CMC): A method of showing measured accuracy performance of a biometric system operating in a closed-set identification task.

D

d **prime (*d'*):** A statistical measure of how well a biometric system can discriminate among various individuals. The larger the *d* prime value, the better a biometric system is at discriminating among individuals.

Data vaulting: The process of sending data off-site where it can be protected from hardware failures, theft, and other threats. It is also referred to as a remote backup service (RBS).

Dead capture: The capture of a biometric sample from biometric evidence.

Decision: The result of the comparison between the score and the threshold. The decisions a biometric system can make include match, nonmatch, and inconclusive, although varying degrees of strong matches and nonmatches are possible.

Decision threshold: The acceptance or rejection of biometric data is dependent on the match score falling above or below the threshold. The threshold is adjustable so that the biometric system can be more or less strict, depending on the requirements of any given biometric application.

Degrees of freedom (DF): It is a statistical measure of the number of independent features on which the precision of a parameter estimate is based.

Detection error trade-off (DET) curve: A graphical plot of decision error rates (e.g., false reject rates vs. false acceptance rates) or matching error rates (false nonmatch rates vs. false match rates).

Difference score: A value returned by a biometric system that indicates the degree of difference between a biometric sample and a biometric reference.

Digital certificate: In the PKI environment, it is the data issued to a user by a certificate authority (CA), which he or she uses during business transactions to prove his or her identity.

Digital signature: The number derived by performing cryptographic operations on the text to be signed. This operation, or hash function (also called hash algorithm), is performed on the binary code of the text. The result is known as the message digest, and it always has a fixed length. The signature algorithm is applied to the message digest, resulting in the digital signature.

Digital signature algorithm (DSA): Presented in 1991 by the NIST and patented in 1993, it is a publicly available one-way algorithm used to generate or verify digital signatures of a text to be signed (not to encrypt/decrypt information). As input, DSA needs (1) the message digest of the message to be signed, (2) the signer's private key, and (3) a random number. Its output is a pair of numbers (often referred to as r and s), which together make up the digital signature. To verify a digital signature, DSA needs as input (1) the message digest of the text to be verified, (2) the signer's public key, and (3) the value s from the signature. DSA then makes a computation, the output of which is called v, for example. If $v = r$, then the signature verifies.

Discriminant training: A means of refining the extraction algorithm so that biometric data from different individuals are as distinct as possible.

E

Encryption: Making information unreadable or difficult to read for unauthorized persons. It is scrambling data so that it becomes very difficult to unscramble or decipher. Scrambled data is called ciphertext, as opposed to unscrambled data, which is called plaintext. Unscrambling ciphertext is called decryption. Data encryption is accomplished through the use of an algorithm and a key. The key is used by the algorithm to scramble and unscramble the data. Encryption does not make unauthorized decryption impossible, but merely difficult. Time and the ever-increasing power of computers are the factors involved in the feasibility of decryption.

End user: A person who interacts with a biometric system to enroll or have his or her identity checked.

End user adaptation: The process of adjustment whereby participants in a test become familiar with what is required and alters their responses accordingly.

Enrollee: A person who has a biometric reference template on file.

Enrollment: The initial process of collecting biometric samples from an individual and the subsequent preparation and storage of biometric reference templates representing that person's identity.

Enrollment template: A biometric template, or digital representation of a physical trait, created during the enrollment process.

Enrollment time: The time period a person must spend to have his or her biometric reference template successfully created.

Equal error rate (EER): A statistic used to measure biometric performance when operating in the verification task. EER occurs when the decision threshold of a system is set so that the proportion of false rejections will be approximately equal to the proportion of false acceptances. EER is a synonym to *crossover rate.*

Extraction: The process of converting a captured biometric sample into biometric data so that it can be compared to a reference template.

F

Face recognition: A biometric modality that uses an image of the physical structure of an individual's face for recognition purposes.

Failure to acquire (FTA): Failure of a biometric system to capture and extract usable biometric data from a biometric sample. The failure to acquire situation might depend on adjustable thresholds for image or signal quality.

Failure to acquire rate (FTAR): The frequency (e.g., percentage of attempts) of a failure to acquire.

Failure to enroll (FTE): Failure of a biometric system to form a proper enrollment reference for an end user.

Failure to enroll rate (FTER): The proportion of a population of users that fail to enroll.

False acceptance: When a biometric system incorrectly identifies an individual or incorrectly verifies an imposter against a claimed identity.

False acceptance rate (FAR): Measures how frequently unauthorized persons are accepted by the system due to erroneous matching. It is the probability that a biometric system will incorrectly verify an individual's identity or will fail to reject an imposter's identity. It is stated as follows: FAR = NFA/NIIA or FAR = NFA/NIVA, where FAR is the false acceptance rate, NFA is the number of false acceptances, NIIA is the number of imposter identification attempts, and NIVA is the number of imposter verification attempts.

False acceptance rate (FAR) curve: Graphic depiction of false acceptances plotted as a function of a decision threshold.

False match rate (FMR): The expected probability that a sample will be falsely declared a match of a single template.

False nonmatch rate (FNMR): The expected probability that a sample will be falsely declared not to match a template from the user supplying the sample.

False rejection: When a biometric system fails to identify an enrollee or fails to verify the legitimate claimed identity of an enrollee.

False rejection rate (FRR): Measures how frequently registered users are rejected by the system. It is the probability that a biometric system will fail to identify an enrollee, or verify the legitimate claimed identity of an enrollee. It is stated as follows: FRR = NFR/NEIA or FRR = NFR/NEVA, where FRR is the false rejection rate, NFR is the number of false rejections, NEIA is the number of enrollee identification attempts, and NEVA is the number of enrollee verification attempts.

False rejection rate (FRR) curve: Graphic depiction of false rejections as a function of the decision threshold.

Feature: A distinctive mathematical characteristic derived from a biometric sample.

Feature extraction: The automated process of locating and encoding distinctive characteristics from a biometric sample in order to generate a template.

Field test: A trial of a biometric application in the real world, as opposed to laboratory conditions.
Filtering: The process of classifying biometric data according to information that is unrelated to the biometric data itself.
Fingerprint recognition: A biometric modality that uses the physical structure of an individual's fingerprint for recognition purposes.
Fusion: The umbrella term for a wide range of methods for the combination of multiple sets of biometric data. These may be sample data, processed data such as templates, matcher similarity scores or distances, verification decisions, and identification candidate lists or ranks. Fusion generically covers the combination of data from multiple samples, multiple (imaging or biometric) modes, or multiple algorithms. Fusion is usually conducted to improve performance.

G

Gait: An individual's manner of walking; a behavioral biometric characteristic.
Gallery: A biometric system database, or set of known individuals, for a specific implementation or evaluation experiment.

H

Hacking: The act of gaining illegal or unauthorized access to a computer system or network.
Hamming distance: The number of noncorresponding digits in a string of binary digits, used to measure dissimilarity.
Hand geometry recognition: A biometric modality that uses the physical structure of an individual's hand for recognition purposes.

I

Identification: The process by which the biometric system identifies a person by performing a one-to-many (1:n) search against the entire enrolled population. Identification systems are designed to determine identity based solely on biometric information.

Identifying characteristics: Unique characteristics of an individual used for biometric processing.

Identity: An assortment of information describing and individual's characteristics and uniqueness. Note: Identity is information concerning an individual, not the individual himself.

Identity credential: Holds information that grants an individual access to predetermined permissions or privileges.

Identity management: The process of establishing identities for individuals in a system and controlling their access to resources with that system by associating user rights and restrictions with the established identities.

Identity management system: A system that is composed of one or more computer systems or applications that manage the identity registration, verification, validation, and issuance process, as well as provisioning and deprovisioning of identity credentials.

Identity proofing: Also called identity vetting, this is the process of providing sufficient information such as background, history, credentials, and documentation that establishes an identity to an organization for the purpose of obtaining identity credentials.

Identity theft: The appropriation of another's personal information to commit fraud, steal the person's assets, or pretend to be that person.

Imposter: A person who submits a biometric sample in either an intentional or inadvertent attempt to pass himself or herself off as another person who is an enrollee.

Integrated Automated Fingerprint Identification System (IAFIS): The Federal Bureau of Investigation's large-scale 10-fingerprint identification system that is used for criminal history background checks and identification of latent prints discovered at crime scenes.

Intraclass variation: A condition where data acquired from a user during verification differs significantly from the data acquired during enrollment by the same user.

Iris recognition: A biometric modality that uses an image of a physical structure of an individual's iris for recognition purposes.

ISO 19794-9; 2007: ISO standard on vein data interchange format.

K

Keystroke dynamics: A biometric modality that uses the cadence of an individual's typing pattern for recognition.

L

Latent fingerprint: These are "leftover" fragments left on a surface that was touched and are usually caused by the buildup of oily residues on a finger.

Live capture: The process of capturing a biometric sample by an interaction between an end user and a biometric system.

Liveness detection: A system determination if a collected biometric is presented by a living person for the purpose of thwarting a spoofing attack.

M

Match: A decision that a biometric sample and a stored biometric template originates from the same individual, based on the high set of similarity difference.

Match score: A numeric value associated with a sample template derived from a comparison to a reference template.

Matching: The comparison of biometric templates to determine their degree of similarity or correlation. In most biometric systems, a match attempt results in a score that is compared against a threshold score. If the matching score exceeds the threshold score, then the result is a match; if the matching score falls below the threshold score, then the result is a nonmatch.

Match-on-card: Storing and matching biometrics on smart cards. The smart card has built-in software that matches the template saved on the card against the input biometric image. As such, the template never leaves the secure environment of the smart card, protecting both the biometric information and the user's personal privacy.

Mifare: An interface for contactless smart cards and smart card readers. It has been developed by Philips and influenced the ISO14443 standard.

Minutiae: The unique, measurable physical characteristics scanned as input and stored for matching by biometric systems. For fingerprints, minutiae points include the starting and ending points of ridges, bifurcations, and ridge junctions among other features.

Minutiae-based method: Minutiae-based algorithms compare several minutiae points (such as ridge endings, bifurcations, or short ridges) extracted from the original image stored in a template with those extracted from a candidate fingerprint. Similar to the pattern-based algorithm, the minutiae-based algorithm must align a fingerprint image before extracting feature points.

Modality: A type or class of biometric systems, for example, vein pattern recognition or hand geometry.

Model: A representation used to characterize an individual; behavioral-based biometrics use models rather than static templates.

Multifactor authentication: The use of multiple techniques to authenticate an individual's identity. This usually involves combining tokens/cards with PINs/passwords and/or biometrics.

Multimodal biometric: A biometric device that uses information from two or more different biometric types, for example, a fingerprint and finger vein.

N

National Institute of Standards and Technology (NIST): An agency of the U.S. federal government within the U.S. Department of Commerce that establishes standards and guidelines for private and public sector purposes.

Near infrared: Light that lies outside the human visible spectrum at its low frequency end.

Noise: Unwanted components in a signal that degrade the quality of data or interfere with the desired signals processed by the system.

Nonrepudiation: The assurance that the authentication of parties to a transaction is so strong that they cannot later deny (repudiate) that they were the parties to that transaction.

Normalization: An algorithm that brings dissimilarly scaled matching scores into a common alignment based on the concept of a normal probability distribution.

O

One-to-many (1:N): Synonym for *identification;* describes the comparison of one sample reference to many enrolled references to make a decision.

One-to-one (1:1): Synonym for *verification;* describes a comparison of one sample reference to one specific enrollment reference to make a decision.

Open-set identification: Situations in which the individual may not be enrolled in the biometric system. This is sometimes referred to as a "watch list" task. The opposite of *closed-set identification.*

Operational evaluation: Used to determine the workflow impact seen by the addition of a biometric system.

Out of set: In open-set identification, an out of set occurs when the individual is not enrolled in the biometric system.

P

Palm print recognition: A biometric modality that uses the physical structure of an individual's palm print for recognition purposes.

Password synchronization: Requires users to remember a single password across different systems, reducing the opportunities to forget one or more of their passwords; however, unlike single sign-on, the user must still enter an ID and password at each application.

Pattern-matching method: Pattern-based algorithms compare the biometric patterns (e.g., for a fingerprint: an arch, whorl, and loop) between a previously stored template and a live sample. This requires that the images be aligned in the same orientation. To do this, the algorithm finds a central point in the image and centers on that. In a pattern-based algorithm, the template contains the type, size, and orientation of patterns within the aligned biometric image.

Performance criteria: Predetermined criteria established to evaluate the performance of the biometric system under testing.

Physical/physiological biometric: A biometric that is characterized by a physical characteristic rather than a behavioral trait.

Population: The set of potential end users for an application.

Presentation: When the user physically presents to the biometric system the data required for capture, such as a finger, hand, or eye.

Privacy: The assurance that data provided or used in a specific transaction will not be revealed or used by the recipient for purposes not authorized by the provider.

Privacy intrusive: A privacy-intrusive system facilitates or enables the usage of personal data in a fashion inconsistent with generally accepted privacy principles.

Privacy protective: A privacy-protective system is one used to protect or limit access to personal information, or which provides a means for an individual to establish a trusted identity.

Privacy neutral: A privacy-neutral system is one in which privacy is not an issue, or in which the potential privacy impact is slight. Privacy-neutral systems are difficult to misuse from a privacy perspective but do not have the capability to protect personal privacy.

Privacy sympathetic: A privacy-sympathetic system is one that limits access to and usage of personal data and in which decisions regarding design issues such as storage and transmission of biometric data are shaped, if not driven, by privacy concerns.

Pruning: Reduces the pattern-matching process by narrowing a large population of potential matches to more manageable subsets.

Pseudonymity: An individual identifier assumed to disguise the true identity of the individual.

Public key infrastructure (PKI): Uses a public/private key, to encrypt IDs, documents, or messages. It starts with a certificate authority (CA), which issues digital certificates.

R

Receiver operating characteristic (ROC) curve: A method of showing measured accuracy performance of a biometric system by comparing graphically for a verification ROC a false accept rate with a verification rate; an open-set identification (watch list) ROC compares false alarm rates to detection and identification rates.

Recognition: A generic term that should not be used in biometrics unless directly associated with specific biometric modality such as speaker recognition or vein pattern recognition.

Reference template: A biometric template for an individual for future comparisons.

Response time: The time period required by a biometric system to return a decision on identification or verification of a biometric sample.

Ridge endings: Minutiae points at the ending of a fracture ridge.

Role-based authorization: A technique of authorization management in which individuals are granted authorization by assignment to one or more predefined roles.

S

Scenario evaluation: Used to measure performance of a biometric system operating in a specific application.

Score: A number indicating the degree of similarity or correlation of a biometric match. Traditional authentication methods—passwords, PINs, keys, and tokens—are binary, offering only a strict yes/no response. However, biometric systems are based on matching algorithms that generate a probability of a match. This probability is converted to a score that represents the degree of correlation between the verification template and the enrollment template. There is no standard scale used for biometric scoring: for some vendors a scale of 1 to 100 might be used, others might use a scale of –1 to 1. Some vendors may use a logarithmic scale and others a linear scale. Regardless of the scale employed this verification score is compared to the system's threshold to determine how successful the potential biometric match is.

Secure identity: The verifiable and exclusive right to use the identity information being presented by an individual to access a set of privileges.

Security: Protection from intended and unintended breaches that would result in the loss or dissemination of data or damage to the integrity, confidentiality, or authenticity of the system.

Segmentation: The process of parsing the biometric signal (item) of interest from the entire acquired data set.

Sensor: The physical hardware device used for biometric capture from cameras to telephones. Sensors convert biometric input into a digitized signal and conveys this information to the processing device.

Signature dynamics: A behavioral biometric modality that analyzes dynamic characteristics of an individual's signature, such as shape of signature, speed of signing, pen pressure, and so forth, for recognition.

Similarity score: A value returned by a biometric algorithm that indicates the degree of similarity or correlation between a biometric sample and a biometric reference.

Single error rates: Error rates state the likelihood of an error (false match, false nonmatch, or failure to enroll) for a single comparison of two biometric templates or for a single enrollment attempt. This can be thought of as a "single" error rate.

Single sign-on (SSO): Enables users to log onto a PC or network and access multiple applications and systems using that single log-on process. Generally, SSO technology requires its own infrastructure; however, it can be biometrically enabled.

Smart card: A device that includes an embedded integrated circuit, memory, and a microcontroller. It can be contact based or contactless.

Soft biometrics: Visual methods for identifying people based on traits that in themselves are not sufficiently distinct but assist in classification of individuals within a database. These traits include gender, eye color, ethnicity, and height.

Speaker recognition: A biometric modality that uses an individual's speech, a feature that is influenced by both an individual's physical and behavioral characteristics, for recognition purposes. It is also called voice verification.

Spoofing: The ability to fool a biometric sensor into recognizing an unauthorized user as a legitimate user (verification) or into missing an identification of someone in the database (identification).

Synchronous multimodality: The use of multiple biometric technologies in a single authentication process. For example, a biometric system might use fingerprints and finger vein pattern recognition simultaneously, reducing the likelihood of fraud and reducing the time needed to verify.

Submission: The process whereby a user provides biometric data in the form of biometric samples to a biometric system. A submission may require looking in the direction of a camera or placing a finger on a reader device. Depending on the biometric system, a user may have to remove eyeglasses, remain still for a number of seconds, or recite a pass phrase in order to provide a biometric sample.

T

Technology evaluation: Measures the performance of biometric systems in general tasks; typically, it reviews only the recognition algorithm component.

Template: A digital representation of biometric data; a reference pattern of a person stored for matching. A biometric template can vary in size from 9 bytes for hand geometry to several thousand bytes for facial recognition.

Template identifier: An input item, such as a PIN, password, or card number, that is used to link a user to his reference template.

Threat: An intentional or unintentional potential event that could compromise the security and integrity of a system.

Threshold: A predefined number or user setting, often controlled by a biometric system administrator, which establishes the minimum degree of correlation necessary for a comparison to be deemed a match. Most thresholds are adjustable to be stricter or less strict.

Throughput rate: The number of end users that a biometric system can process within a stated time interval.

Token: A physical device containing individual credentials that an authorized person carries to aid in authentication. Hardware tokens are often small enough to be carried in a pocket or purse. Some may store cryptographic keys, like a digital signature, or biometric template.

V

Validation: The process of demonstrating that the system under consideration meets in all respects the specification of that system.

Verification: It is the process of establishing the validity of a claimed identity by comparing a verification template to an enrollment template. Verification requires that an identity be claimed, after which the individual's enrollment template is located and compared with the verification template. Verification answers the question, "Am I who I claim to be?" It is used interchangeably with *authentification*.

Vulnerability: The potential for a biometric system to be compromised by intent (e.g., fraudulent activity), by design flaw (including usage errors), by accident, by hardware malfunctions, or external environmental conditions.

W

Watch list: Refers to an open-set identification; a term to describe answers to the questions: Is this person in the database? If so, who is he?

Whorl: A fingerprint pattern in which the ridges are circular or nearly circular.

X

X.509: A standard that defines what information can go into a certificate and describes how to write it down (the data format).

INDEX

"f" indicates material in figures. "t" indicates material in tables. "n" indicates material in footnotes.

A

B

C

D

E

I

L

M

N

Q

R

S

U

V

W

For Product Safety Concerns and Information please contact our EU representative GPSR@taylorandfrancis.com Taylor & Francis Verlag GmbH, Kaufingerstraße 24, 80331 München, Germany

Printed and bound by CPI Group (UK) Ltd, Croydon, CR0 4YY
01/05/2025
01858353-0002